U.S. Department
of Transportation

National Highway
Traffic Safety
Administration

DOT-VNTSC-NHTSA-02-04
DOT HS 809 573

February 2003

Analysis of Light Vehicle Crashes and Pre-Crash Scenarios Based on the 2000 General Estimates System

Research and
Special Programs
Administration
Volpe National
Transportation Systems Center
Cambridge, MA 02142-1093

This document is available to the public through the National Technical Information Service, Springfield, VA 22161

NOTICE

This document is disseminated under the sponsorship of the Department of Transportation in the interest of information exchange. The United States Government assumes no liability for its contents or use thereof.

REPORT DOCUMENTATION PAGE

Form Approved
OMB No. 0704-0188

Public reporting burden for this collection of information is estimated to average 1 hour per response, including the time for reviewing instructions, searching existing data sources, gathering and maintaining the data needed, and completing and reviewing the collection of information. Send comments regarding this burden estimate or any other aspect of this collection of information, including suggestions for reducing this burden, to Washington Headquarters Services, Directorate for Information Operations and Reports, 1215 Jefferson Davis Highway, Suite 1204, Arlington, VA 22202-4302, and to the Office of Management and Budget, Paperwork Reduction Project (0704-0188), Washington, DC 20503.

1. AGENCY USE ONLY (Leave blank)	2. REPORT DATE February 2003	3. REPORT TYPE AND DATES COVERED Final Report June 2001 – February 2003
4. TITLE AND SUBTITLE Analysis of Light Vehicle Crashes and Pre-Crash Scenarios Based on the 2000 General Estimates System		5. FUNDING NUMBERS HS319/S3059
6. AUTHOR(S) Wassim G. Najm, Basav Sen,* John D. Smith, and Brittany N. Campbell		
7. PERFORMING ORGANIZATION NAME(S) AND ADDRESS(ES) U.S. Department of Transportation Research and Special Programs Administration John A. Volpe National Transportation Systems Center Cambridge, MA 02142		8. PERFORMING ORGANIZATION REPORT NUMBER DOT-VNTSC-NHTSA-02-04
9. SPONSORING/MONITORING AGENCY NAME(S) AND ADDRESS(ES) U.S. Department of Transportation National Highway Traffic Safety Administration 400 7th St. SW Washington, DC 20590		10. SPONSORING/MONITORING AGENCY REPORT NUMBER DOT HS 809 573

11. SUPPLEMENTARY NOTES
*EG&G Technical Services, Inc.
55 Broadway
Cambridge, MA 02142

12a. DISTRIBUTION/AVAILABILITY STATEMENT This document is available to the public through the National Technical Information Service, Springfield, Virginia 22161.	12b. DISTRIBUTION CODE

13. ABSTRACT (Maximum 200 words)

This report analyzes the problem of light vehicle crashes in the United States to support the development and assessment of effective crash avoidance systems as part of the U.S. Department of Transportation's Intelligent Vehicle Initiative. The analysis was conducted using data from the 2000 National Automotive Sampling/General Estimates System crash database of the National Highway Traffic Safety Administration. Light vehicle (passenger cars, sport utility vehicles, vans, and pickup trucks) crashes are analyzed in terms of their major crash types, physical setting, and concomitant pre-crash scenarios. In 2000, light vehicle crashes accounted for 6,133,000 or 96 percent of all police-reported (PR) crashes on U.S. roadways. About 96 percent of all PR crashes belong to nine known major crash types: rear-end, crossing paths, off-roadway, lane change, opposite direction, pedestrian, pedalcyclist, animal, and backing. The examination of the physical setting of major crash types shows that about 40 percent of all PR light vehicle crashes happened away from junctions, 25 percent of all PR light vehicle crashes were reported to occur at intersections, and 20 percent of the crashes were related to intersections. The 9 major crash types consist mainly of 55 specific and dominant pre-crash scenarios. These scenarios yielded a top 11 list of major pre-crash scenarios, which represent a new crash taxonomy that covers 4,275,000 (70 percent) of all PR light vehicle crashes.

14. SUBJECT TERMS light vehicle, crashes, crash-imminent scenarios, test scenarios, Intelligent Vehicle Initiative			15. NUMBER OF PAGES 76
			16. PRICE CODE
17. SECURITY CLASSIFICATION OF REPORT Unclassified	18. SECURITY CLASSIFICATION OF THIS PAGE Unclassified	19. SECURITY CLASSIFICATION OF ABSTRACT Unclassified	20. LIMITATION OF ABSTRACT

PREFACE

The National Highway Traffic Safety Administration (NHTSA), in conjunction with the Research and Special Programs Administration Volpe National Transportation Systems Center (Volpe Center), is conducting an analysis of light vehicle crashes in support of the U.S. Department of Transportation's Intelligent Vehicle Initiative (IVI). The IVI focuses on solving traffic safety problems through the development and deployment of vehicle-based and vehicle-infrastructure cooperative crash countermeasures that address rear-end, roadway departure, lane change, crossing paths, driver impairment, reduced visibility, vehicle instability, pedestrian, and pedalcyclist crashes. Research is being performed in the context of four vehicle platforms including light vehicles (passenger cars, sport utility vehicles, vans, and pickup trucks), commercial vehicles (medium and heavy trucks), transit vehicles (buses, but not school buses), and specialty vehicles (police, fire, ambulance, snow plows, and other roadway maintenance vehicles).

This report presents the results obtained from the analysis of light vehicle crashes using the 2000 National Automotive Sampling System/General Estimates System crash database. In 2000, there were an estimated 6,394,000 police-reported (PR) motor vehicle crashes in the United States that resulted in 41,821 fatalities and 3,189,000 injured people. Light vehicle crashes accounted for 6,133,000 or 96 percent of all PR crashes on U.S. roadways.

The authors of this report are Wassim G. Najm, John D. Smith, and Brittany N. Campbell of the Volpe Center, and Basav Sen of EG&G Technical Services, Inc.

The authors acknowledge the technical contribution of Dr. David L. Smith of NHTSA. Also acknowledged are Dan Cohen of Mitretek and NHTSA staffs from various offices for reviewing the report and providing valuable comments. Kate Klotz of Planners Collaborative edited the report.

METRIC/ENGLISH CONVERSION FACTORS

ENGLISH TO METRIC

LENGTH (APPROXIMATE)
- 1 inch (in) = 2.5 centimeters (cm)
- 1 foot (ft) = 30 centimeters (cm)
- 1 yard (yd) = 0.9 meter (m)
- 1 mile (mi) = 1.6 kilometers (km)

AREA (APPROXIMATE)
- 1 square inch (sq in, in^2) = 6.5 square centimeters (cm^2)
- 1 square foot (sq ft, ft^2) = 0.09 square meter (m^2)
- 1 square yard (sq yd, yd^2) = 0.8 square meter (m^2)
- 1 square mile (sq mi, mi^2) = 2.6 square kilometers (km^2)
- 1 acre = 0.4 hectare (he) = 4,000 square meters (m^2)

MASS - WEIGHT (APPROXIMATE)
- 1 ounce (oz) = 28 grams (gm)
- 1 pound (lb) = 0.45 kilogram (kg)
- 1 short ton = 2,000 pounds (lb) = 0.9 tonne (t)

VOLUME (APPROXIMATE)
- 1 teaspoon (tsp) = 5 milliliters (ml)
- 1 tablespoon (tbsp) = 15 milliliters (ml)
- 1 fluid ounce (fl oz) = 30 milliliters (ml)
- 1 cup (c) = 0.24 liter (l)
- 1 pint (pt) = 0.47 liter (l)
- 1 quart (qt) = 0.96 liter (l)
- 1 gallon (gal) = 3.8 liters (l)
- 1 cubic foot (cu ft, ft^3) = 0.03 cubic meter (m^3)
- 1 cubic yard (cu yd, yd^3) = 0.76 cubic meter (m^3)

TEMPERATURE (EXACT)
[(x-32)(5/9)] °F = y °C

METRIC TO ENGLISH

LENGTH (APPROXIMATE)
- 1 millimeter (mm) = 0.04 inch (in)
- 1 centimeter (cm) = 0.4 inch (in)
- 1 meter (m) = 3.3 feet (ft)
- 1 meter (m) = 1.1 yards (yd)
- 1 kilometer (km) = 0.6 mile (mi)

AREA (APPROXIMATE)
- 1 square centimeter (cm^2) = 0.16 square inch (sq in, in^2)
- 1 square meter (m^2) = 1.2 square yards (sq yd, yd^2)
- 1 square kilometer (km^2) = 0.4 square mile (sq mi, mi^2)
- 10,000 square meters (m^2) = 1 hectare (ha) = 2.5 acres

MASS - WEIGHT (APPROXIMATE)
- 1 gram (gm) = 0.036 ounce (oz)
- 1 kilogram (kg) = 2.2 pounds (lb)
- 1 tonne (t) = 1,000 kilograms (kg)
- = 1.1 short tons

VOLUME (APPROXIMATE)
- 1 milliliter (ml) = 0.03 fluid ounce (fl oz)
- 1 liter (l) = 2.1 pints (pt)
- 1 liter (l) = 1.06 quarts (qt)
- 1 liter (l) = 0.26 gallon (gal)
- 1 cubic meter (m^3) = 36 cubic feet (cu ft, ft^3)
- 1 cubic meter (m^3) = 1.3 cubic yards (cu yd, yd^3)

TEMPERATURE (EXACT)
[(9/5) y + 32] °C = x °F

QUICK INCH - CENTIMETER LENGTH CONVERSION

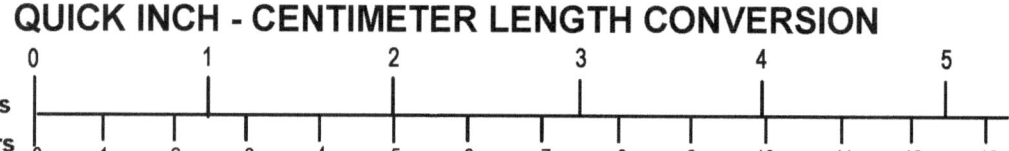

QUICK FAHRENHEIT - CELSIUS TEMPERATURE CONVERSION

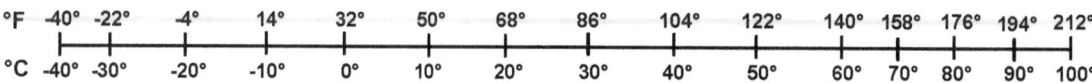

For more exact and or other conversion factors, see NIST Miscellaneous Publication 286, Units of Weights and Measures. Price $2.50 SD Catalog No. C13 10286

Updated 6/17/98

TABLE OF CONTENTS

Section	Page
Executive Summary	x
1. Introduction	1
1.1 Previous Research	2
1.2 Report Outline	3
2. Light Vehicle Crash Types and Pre-Crash Scenarios	5
2.1 Distribution of Crash Types for All Vehicles and Light Vehicles	5
2.2 Rear-End Crashes and Pre-Crash Scenarios	8
2.3 Crossing Path Crashes and Pre-Crash Scenarios	9
2.4 Off-Roadway Crashes and Pre-Crash Scenarios	12
2.5 Lane Change Crashes and Pre-Crash Scenarios	15
2.6 Animal Crashes and Pre-Crash Scenarios	17
2.7 Opposite Direction Crashes and Pre-Crash Scenarios	18
2.8 Backing Crashes and Pre-Crash Scenarios	20
2.9 Pedestrian Crashes and Pre-Crash Scenarios	21
2.10 Pedalcyclist Crashes and Pre-Crash Scenarios	23
3. Physical Setting of Crashes and Pre-Crash Scenarios	25
3.1 Rear-End Crashes	27
3.2 Crossing Path Crashes	31
3.3 Off-Roadway Crashes	32
3.4 Lane Change Crashes	34
3.5 Animal Crashes	37
3.6 Opposite Direction Crashes	39
3.7 Pedestrian Crashes	41
3.8 Pedalcyclist Crashes	42
3.9 Light Vehicle Crash Distribution by Traffic Control Device	45
4. Crash Taxonomy	47
4.1 Development of Pre-Crash Scenario Taxonomy	47
4.2 Major Pre-Crash Scenarios	49
4.2.1 Lead Vehicle Stopped	50
4.2.2 Straight Crossing Paths	50
4.2.3 Control Loss	50
4.2.4 Left Turn Across Path/Opposite Direction	51
4.2.5 Drifting (Going Straight)	51
4.2.6 Lead Vehicle Decelerating	51
4.2.7 Left Turn Across Path/Lateral Direction	52
4.2.8 Simple Lane Change	52
4.2.9 Vehicle Going Straight/Animal in Roadway	52
4.2.10 Drifting (Negotiating a Curve)	52
4.2.11 Lead Vehicle Moving at Constant Speed	52

TABLE OF CONTENTS (cont.)

Section	**Page**
5. Conclusion	55
References	57
Appendix A. 2000 General Estimates System Estimates and Standard Errors	59
Appendix B. General Estimates System Analysis Codes	61

LIST OF FIGURES

Figure   **Page**

2-1(a). Share of Crashes by Type for All Vehicles 7
2-1(b). Share of Crashes by Type for Light Vehicles 7
2-2. Distribution of Pre-Crash Scenarios for Rear-End Crashes of Light Vehicles 9
2-3. Distribution of Pre-Crash Scenarios for Crossing Path Crashes of Light Vehicles 10
2-4. Schematics of Common Crossing Path Pre-Crash Scenarios 11
2-5(a). Distribution of Crash Subtypes for Off-Roadway Crashes of Light Vehicles 14
2-5(b). Distribution of Pre-Crash Scenarios for Single Vehicle Off-Roadway Crashes of Light Vehicles 14
2-6. Distribution of Major Pre-Crash Scenarios for Lane Change Crashes of Light Vehicles 17
2-7. Distribution of Pre-Crash Scenarios for Crashes of Light Vehicles with Animals 18
2-8. Distribution of Pre-Crash Scenarios for Opposite Direction Crashes of Light Vehicles 20
2-9. Distribution of Pre-Crash Scenarios for Backing Crashes of Light Vehicles 21
2-10. Distribution of Pre-Crash Scenarios for Crashes of Light Vehicles with Pedestrians 23
2-11. Distribution of Pre-Crash Scenarios for Crashes of Light Vehicles with Pedalcyclists 24
3-1(a). Distribution of All Vehicle Crashes by Relation to Junction 26
3-1(b). Distribution of Light Vehicle Crashes by Relation to Junction 27
3-2(a). Distribution of Rear-end Crashes by Relation to Junction 29
3-2(b). Distribution of Lead Vehicle Stopped Pre-Crash Scenario for Rear-end Crashes of Light Vehicles by Relation to Junction 29
3-2(c). Distribution of Lead Vehicle Decelerating Pre-Crash Scenario for Rear-end Crashes of Light Vehicles by Relation to Junction 30
3-3. Distribution of Crossing Path Crashes of Light Vehicles by Relation to Junction 32
3-4. Distribution of Crash Subtypes for Off-Roadway Crashes of Light Vehicles by Relation to Junction 33
3-5(a). Distribution of Lane Change Crashes of Light Vehicles by Relation to Junction ... 36
3-5(b). Distribution of Pre-Crash Scenario 1 (One Vehicle Going Straight, One Vehicle Executing Simple Lane Change) for Lane Change Crashes of Light Vehicles by Relation to Junction 36
3-5(c). Distribution of Pre-Crash Scenario 2 (One Vehicle Going Straight, One Vehicle Turning) for Lane Change Crashes of Light Vehicles by Relation to Junction 37
3-6. Distribution of Crashes of Light Vehicles with Animals by Relation to Junction 38
3-7(a). Distribution of Opposite Direction Crashes of Light Vehicles by Relation to Junction ... 40
3-7(b). Distribution of Going Straight/Encroaching Pre-Crash Scenario for Opposite Direction Crashes of Light Vehicles by Relation to Junction 40
3-7(c). Distribution of Negotiating a Curve/Encroaching Pre-Crash Scenario for Opposite Direction Crashes of Light Vehicles by Relation to Junction 41
3-8. Distribution of Pedestrian Crashes of Light Vehicles by Relation to Junction 42
3-9(a). Distribution of Pedalcyclist Crashes of Light Vehicles by Relation to Junction 44
3-9(b). Distribution of Vehicle Turning Left/Parallel Paths Pre-Crash Scenario for Pedalcyclist Crashes of Light Vehicles by Relation to Junction 44

LIST OF TABLES

Table   **Page**

2-1. Comparison of Numbers of Vehicles and VMT for Light Vehicles and All Vehicles 5
2.2(a). Shares of PR Crashes by Type for All Vehicles 6
2.2(b). Shares of PR Crashes by Type for Light Vehicles 6
2-3. Distribution of Pre-Crash Scenarios for Rear-End Crashes of Light Vehicles 8
2-4. Distribution of Pre-Crash Scenarios for Crossing Path Crashes of Light Vehicles 10
2-5(a). Distribution of Crash Subtypes for Off-Roadway Crashes of Light Vehicles 13
2-5(b). Distribution of Pre-Crash Scenarios for Single Vehicle Off-Roadway Crashes of Light Vehicles 13
2-6(a). Pre-Crash Scenario Matrix for Lane Change Crashes of Light Vehicles 16
2-6(b). Distribution of Major Pre-Crash Scenarios for Lane Change Crashes of Light Vehicles 16
2-7. Distribution of Pre-Crash Scenarios for Crashes of Light Vehicles with Animals 18
2-8. Distribution of Pre-Crash Scenarios for Opposite Direction Crashes of Light Vehicles 19
2-9. Distribution of Pre-Crash Scenarios for Backing Crashes of Light Vehicles 21
2-10. Distribution of Pre-Crash Scenarios for Crashes of Light Vehicles with Pedestrians 22
2-11. Distribution of Pre-Crash Scenarios for Crashes of Light Vehicles with Pedalcyclists 23
3-1(a). Distribution of All Vehicle Crashes by Relation to Junction 25
3-1(b). Distribution of Light Vehicle Crashes by Relation to Junction 26
3-2. Distribution of Pre-Crash Scenarios for Rear-End Crashes of Light Vehicles by Relation to Junction 28
3-3. Distribution of Pre-Crash Scenarios for Crossing Path Crashes of Light Vehicles by Relation to Junction 31
3-4. Distribution of Crash Subtypes for Off-Roadway Crashes of Light Vehicles by Relation to Junction 33
3-5. Distribution of Pre-Crash Scenarios for Lane Change Crashes of Light Vehicles by Relation to Junction 35
3-6. Distribution of Pre-Crash Scenarios for Crashes of Light Vehicles with Animals by Relation to Junction 38
3-7. Distribution of Pre-Crash Scenarios for Opposite Direction Crashes of Light Vehicles by Relation to Junction 39
3-8. Distribution of Pre-Crash Scenarios for Pedestrian Crashes of Light Vehicles by Relation to Junction 42
3-9. Distribution of Pre-Crash Scenarios for Pedalcyclist Crashes of Light Vehicles by Relation to Junction 43
3-10. Distribution of PR Light Vehicle Crashes by Relation to Junction by Traffic Control Device 45
3-11. Distribution of Crash Type by Relation to Junction by Traffic Control Device 46
4-1. Distribution of Top Pre-Crash Scenarios for Light Vehicle Crashes by Crash Type 48

LIST OF ACRONYMS

CARDfile	Crash Avoidance Research Data File
CL	Control Loss
DOT	Department of Transportation
GES	General Estimates System
ITS	Intelligent Transportation Systems
IVI	Intelligent Vehicle Initiative
LD	Lateral Direction
LTAP	Left Turn Across Path
LTIP	Left Turn Into Path
LVD	Lead Vehicle Decelerating
LVS	Lead Vehicle Stopped
NHTSA	National Highway Traffic Safety Administration
OD	Opposite Direction
PR	Police Reported
RTAP	Right Turn Across Path
RTIP	Right Turn Into Path
SCP	Straight Crossing Paths

EXECUTIVE SUMMARY

This report analyzes the problem of light vehicle crashes in the United States to support the development and assessment of effective crash avoidance systems as part of the U.S. Department of Transportation's Intelligent Vehicle Initiative. Light vehicle (passenger cars, sport utility vehicles, vans, and pickup trucks) crashes are analyzed in terms of their major crash types, physical setting, and concomitant pre-crash scenarios. In 2000, light vehicle crashes accounted for 6,133,000 or 96 percent of all police-reported (PR) crashes on U.S. roadways. The analysis was conducted using data from the 2000 National Automotive Sampling System/General Estimates System (NASS/GES) crash database of the National Highway Traffic Safety Administration.

About 96 percent of all PR light vehicle crashes belong to nine known major crash types: rear-end, crossing paths, off-roadway, lane change, opposite direction, pedestrian, pedalcyclist, animal, and backing. The first four crash types dominate the population of PR light vehicle crashes with a combined frequency of 5,241,000 or 85 percent of all these crashes. The examination of the physical setting shows that about 40 percent of all PR light vehicle crashes happened away from junctions. Next to non-junctions, 24.5 percent and 20.4 percent of all PR light vehicle crashes were reported respectively at intersections or related to intersections (i.e., on roadways close to and leading to intersections). The majority of off-roadway crashes (74.5 percent), opposite direction crashes (81.0 percent), and animal crashes (95.6 percent) happened away from junctions. Moreover, pedestrian crashes as well as lane change crashes were reported more at non-junctions than at any other location. As expected, most crossing path crashes (73.7 percent) occurred within the confines of intersections. Unlike pedestrian crashes, more pedalcyclist crashes occurred at intersections than at any other location. About 44 percent of all rear-end crashes were coded as intersection-related crashes. Driveways were the most reported location for backing crashes (≈ 39 percent), followed by intersection-related roadways (≈ 30 percent). The presence of traffic signals was mostly reported in pedestrian, rear-end, lane change, crossing paths, and backing crashes. On the other hand, the stop sign was the most dominant traffic device for pedalcyclist crashes.

This report identifies pre-crash scenarios of the nine major crash types based primarily on the analysis of the NASS/GES *Accident Type*, *Movement Prior to Critical Event*, and *Critical Event* variables. By definition, pre-crash scenarios combine vehicle movements and maneuvers with critical events that occur immediately prior to a collision. Collectively, the nine major crash types consist mainly of fifty-six specific and dominant pre-crash scenarios. These scenarios are not mutually independent since some scenarios in one major crash type are also reported to happen immediately prior to other crash types. A cross cutting analysis of scenarios yielded a top 11 list of major pre-crash scenarios with individual frequency of at least 100,000 PR crashes:

1. Lead vehicle stopped (888,000)
2. Straight crossing paths (557,000)
3. Control loss (486,000)
4. Left turn across path/opposite direction (424,000)
5. Drifting (going straight) (400,000)
6. Lead vehicle decelerating (397,000)

7. Left turn across path/lateral direction (304,000)
8. Vehicle changing lanes (274,000)
9. Vehicle going straight/animal in roadway (232,000)
10. Drifting (negotiating a curve) (174,000)
11. Lead vehicle moving at constant speed (139,000)

This list represents a new crash taxonomy that covers 4,275,000 or 70 percent of all PR light vehicle crashes. Finally, this new crash taxonomy highlights the significant contribution of vehicle drifting and control loss (due to excessive speeding) to light vehicle crashes.

1. INTRODUCTION

This report provides a comprehensive analysis of all police-reported (PR) crashes that involved light vehicles (passenger cars, sport utility vehicles, vans, and pickup trucks) based on statistics from the 2000 National Automotive Sampling System/General Estimates System (GES) crash database. This analysis describes light vehicle crashes in terms of their major crash types, physical setting, and concomitant pre-crash scenarios. In 2000, there were an estimated 6,394,000 PR motor vehicle crashes in the United States, which resulted in 41,821 fatalities and 3,189,000 injured people [1]. Approximately 67 percent of these crashes involved property damage only. Light vehicle crashes accounted for 6,133,000 or 96 percent of all police-reported crashes on U.S. roadways.

This analysis of light vehicle crashes was conducted in support of the U.S. Department of Transportation's (DOT) Intelligent Transportation Systems (ITS)/Intelligent Vehicle Initiative (IVI). The goal of the IVI is to facilitate the development and accelerate the deployment of advanced-technology crash avoidance systems that significantly improve the collision avoidance capabilities of motor vehicles [2]. The IVI emphasizes the significant and continuing role of the driver in solving traffic safety problems by means of effective vehicle-based safety systems [3]. Research is being performed in the context of four vehicle platforms including light vehicles, commercial vehicles (medium and heavy trucks), transit vehicles (buses, but not school buses), and specialty vehicles (police, fire, ambulance, snow plows, and other roadway maintenance vehicles). The major part of this report concentrates on light vehicles only, though a brief discussion of crashes of all vehicles is provided for context.

This report defines a typology of prevalent light vehicle crash types, describes their physical setting in terms of crash relation to roadway junctions, and identifies concomitant pre-crash scenarios based on the 2000 GES. Moreover, this report transforms a classification of major crash types into taxonomy of most common pre-crash scenarios that may precede one or more crash types. Pre-crash scenarios refer to vehicle movements and critical events that occur immediately prior to collision. The characterization of the sequence of events leading to collisions is essential to the design of appropriate crash countermeasure systems, the development of their performance specifications and objective test procedures, and the estimation of their safety benefits. The combination of pre-crash scenarios and causal factors allows the development of crash countermeasure concepts and essential functional requirements [4]. Information on pre-crash scenarios and their physical setting helps to develop performance guidelines and objective test procedures, and guides researchers to collect the appropriate data on driver performance with and without the assistance of crash avoidance systems [5]. Such data are essential to the design of effective warning algorithms and driver-vehicle interfaces, as well as the estimation of safety benefits for crash avoidance systems [6].

The National Highway Traffic Safety Administration's (NHTSA) GES crash database is generally used to identify highway safety problem areas, supply a foundation for regulatory and consumer information initiatives, and form the basis for cost and benefit analyses of highway safety initiatives [7]. The GES is a nationally representative sample of PR crashes in the United States, collected from about 400 police agencies within 60 geographical sites. About 55,000 police accident reports are selected each year and coded directly in the GES by trained personnel who check the data for validity and consistency. Although various sources suggest that about half the motor vehicle crashes in the country are not reported to the police, the majority of these unreported crashes involve only minor property damage and no significant personal injury. By restricting attention to PR crashes, the GES concentrates on those crashes of greatest concern to the highway safety community and the general public.

1.1. Previous Research

An early ITS study defined a typology of prevalent crash types to identify and evaluate the application of new infrastructure-based technology to known highway safety problems, including an assessment of functional requirements, feasibility, costs, and potential safety benefits [8]. Six crash types that involved all motor vehicles were selected as targets for the application of advanced technology based on an analysis of NHTSA's 1984-1986 Crash Avoidance Research Data file (CARDfile). The six types include the following single-vehicle and two-vehicle crash types:

1. Run-off-road crash: single vehicle strikes a fixed object or overturns off the roadway.
2. Single vehicle strikes pedestrian, cyclist, or animal.
3. Crossing paths at intersection or driveway: two vehicles both going straight.
4. Left-turn crash: one vehicle turns left across path of another at intersection or driveway.
5. Rear-end crash between two vehicles moving in the same direction.
6. Head-on crash between two vehicles approaching from opposite directions.

Another study examined 12 major types and subtypes of all motor vehicle crashes based on 1993 GES statistics to define ITS collision avoidance system concepts [4]. This crash typology consists of the following types and subtypes:

1. Rear-end crash:
    a. Lead vehicle stationary.
    b. Lead vehicle moving.
2. Backing crash:
    a. Encroachment: slowly moving backing vehicle strikes pedestrian, object, or vehicle.
    b. Crossing paths: backing vehicle (e.g., out of a driveway) collides with a moving vehicle (e.g., traveling "at speed" on a street).
3. Lane change/merge crash:
    a. Angle/sideswipe crash between two vehicles in adjacent lanes moving in the same direction.
    b. Rear-end crash: lane changing/merging vehicle is rear-ended after the lane change/merge maneuver.
4. Single vehicle roadway departure crash.
5. Intersection crossing path crash:
    a. Signalized intersection straight crossing paths.
    b. Unsignalized intersection straight crossing paths.
    c. Left turn across path.

   d. Other intersection crossing paths.
6. Opposite direction crash: two vehicles moving in opposite directions collide head-on, at an angle, or sideswipe.

Later studies utilized the GES pre-crash variables to better describe vehicle movements and critical events prior to impact. The identification of such pre-crash scenarios was primarily performed by a query of two GES pre-crash variables. The first *Movement Prior to Critical Event* pre-crash variable describes a vehicle's activity prior to the driver's realization of an impending critical event or danger. This variable discerns vehicle maneuvers, such as passing or turning, and dynamic states such as stopped or decelerating. The second *Critical Event* pre-crash variable identifies the critical event that made the crash imminent. The results of these studies are as follows:

- Ten prevalent pre-crash scenarios were identified from 1992-1996 GES statistics on rear-end crashes of all vehicles [9].
- Crossing path crashes were divided into six common pre-crash scenarios for all vehicles and light vehicles based on 1998 GES data [10].
- Single light vehicle off-roadway crashes were grouped into six pre-crash scenarios based on vehicle movements (going straight, negotiating a curve, or initiating a maneuver) and critical events (departed roadway edge or lost control) [11].
- Pedestrian crashes with all vehicles were separated into nine basic pre-crash scenarios using 1995-1998 GES data [12].
- A breakdown of pedalcyclist crashes with all vehicles revealed a total of 8 pre-crash scenarios based on 1995-1998 GES data [13].
- An analysis of the 1999 GES data identified seven of the most common pre-crash scenarios in all and light vehicle lane change crashes including lane changing, turning, drifting, passing, parking, and merging maneuvers [14].

1.2. Report Outline

Following the introduction, Section 2 describes the classification of light vehicle crashes into major crash types and identifies their pre-crash scenarios. Section 3 provides the frequency distribution of each crash type, as well as major pre-crash scenarios in each crash type, by relation to junction and the presence of traffic signals at intersections. Section 4 discusses the transformation of the classification of crash types into taxonomy of pre-crash scenarios, identifies crosscutting scenarios, ranks scenarios by importance in terms of frequency of occurrence, and provides a more detailed discussion of a selection of major scenarios. The conclusion of this report is presented in Section 5. Finally, this report provides the standard errors of the 2000 GES estimates in Appendix A and delineates the GES codes used for this analysis in Appendix B.

2. LIGHT VEHICLE CRASH TYPES AND PRE-CRASH SCENARIOS

This section describes the major crash types for light vehicles and the common pre-crash scenarios for each type. To provide a context, it also discusses the crash type distribution for all vehicle types.

To obtain a perspective on the relative importance of light vehicles as a share of all vehicular traffic in the United States, the total numbers of registered vehicles in the fleet and the total vehicle miles traveled (VMT) in millions for light vehicles and for all vehicles are compared in Table 2-1. Light vehicles comprise 94 percent of all vehicles in operation and accumulate 92 percent of all VMT on U.S. roadways. Hence, the distributions shown for light vehicles in the remainder of this section are likely to be substantially similar to the distributions for all vehicles.

Table 2-1. Comparison of Numbers of Vehicles and VMT for Light Vehicles and All Vehicles (Based on Year 2000 Statistics)

Vehicle Type	Number of Vehicles	VMT (Millions)	VMT per Vehicle (thousands)
Light vehicles[1]	212,706,399	2,525,932	11.9
All vehicles	225,821,241	2,749,803	12.2
Light vehicles as % of all vehicles	94.2%	91.9%	

Source: Table VM-1 in *Highway Statistics 2000* (USDOT/FHWA, 2001).
Notes: [1] Defined as "Passenger Cars and Other 2-Axle 4-Tire Vehicles" in Table VM-1.

2.1. Distribution of Crash Types for All Vehicles and Light Vehicles

In 2000, there were approximately 6,389,000 PR motor vehicle crashes based on 2000 GES estimates. Of these, as many as 6,133,000 or 96 percent involved at least 1 light vehicle. The dominance of light vehicle crashes is expected based on the shares of number of vehicles and VMT shown in Table 2-1. The national estimates produced from GES data may differ from the true values, because they are based on a probability sample of crashes and not a census of all crashes. The size of these differences may vary depending on which sample of crashes was selected. Generalized standard errors for 2000 GES estimates of totals are provided in Appendix A [7].

The dominance of light vehicles also leads to the distribution of crashes by type for all vehicles (Table 2-2(a) and Figure 2-1(a)) to be almost identical to the distribution of crashes by type for light vehicles (Table 2-2(b) and Figure 2-1(b)). The crash types are described in greater detail in subsequent sections. Appendix B provides the GES codes used to identify these crash types.

Table 2-2(a). Share of PR Crashes by Type for All Vehicles

Crash Type	Number of Crashes	Share of Crashes by Type
Rear-End	1,816,000	28.4%
Crossing Paths	1,594,000	24.9%
Off Roadway	1,448,000	22.7%
Lane Change	572,000	9.0%
Animal	256,000	4.0%
Opposite Direction	168,000	2.6%
Backing	131,000	2.1%
Pedestrian	73,000	1.1%
Pedalcyclist	51,000	0.8%
Other	280,000	4.4%
Total	**6,389,000**	**100.0%**

Note: "Other" crashes are the sum of crashes with objects, undefined crashes, and other crashes not classified as any of the above types.

Table 2-2(b). Share of PR Crashes by Type for Light Vehicles

Crash Type	Number of Crashes	Share of Crashes by Type
Rear-End	1,806,000	29.4%
Crossing Paths	1,590,000	25.9%
Off Roadway	1,280,000	20.9%
Lane Change	565,000	9.2%
Animal	247,000	4.0%
Opposite Direction	163,000	2.7%
Backing	129,000	2.1%
Pedestrian	66,000	1.1%
Pedalcyclist	47,000	0.8%
Other	240,000	3.9%
Total	**6,133,000**	**100.0%**

Note: "Other" crashes are the sum of crashes with objects, undefined crashes, and other crashes not classified as any of the above types.

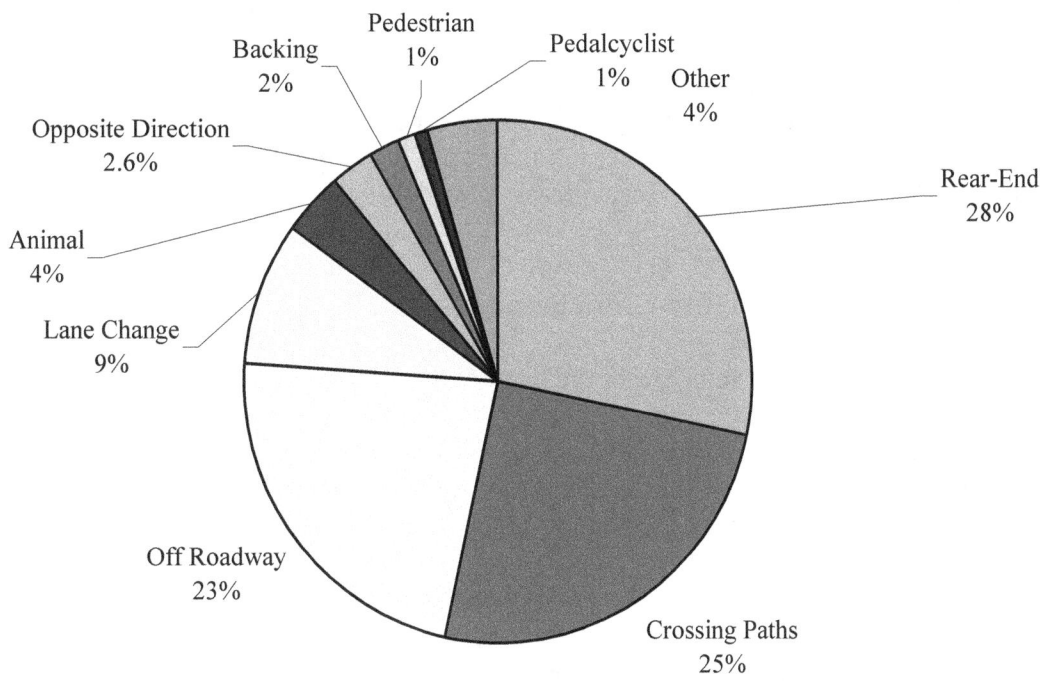

Figure 2-1(a). Share of Crashes by Type for All Vehicles

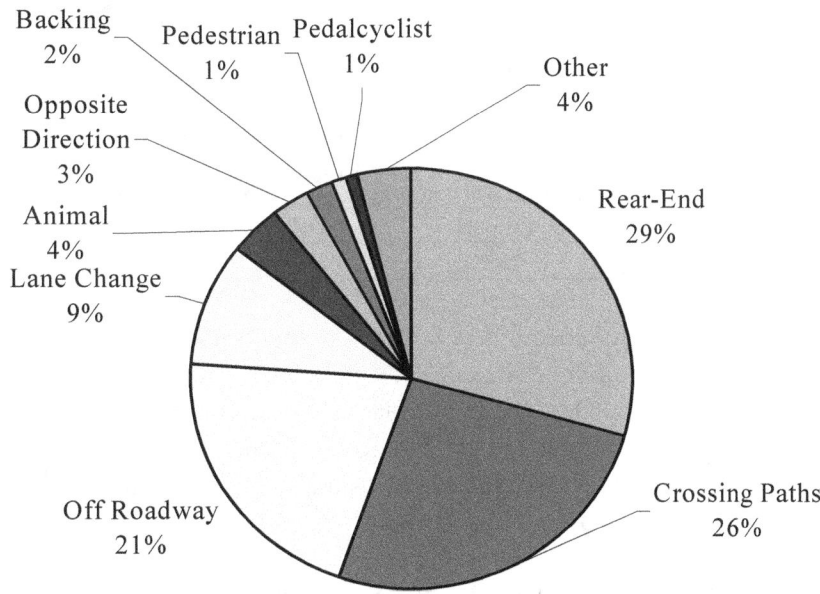

Figure 2-1(b). Share of Crashes by Type for Light Vehicles

Rear-end crashes have the most frequency of occurrence among other crash types, accounting for more than a quarter of the crashes. Crossing path crashes follow with about a quarter of all crashes. Off-roadway and lane change crashes represent the other two dominant crash types. Combined, these 4 crash types account for about 85 percent of all crash types.

The following sections examine pre-crash scenarios for each crash type, *focusing only on light vehicle crashes*. Unless otherwise stated, only single- or two-vehicle crashes are considered, since crashes involving more than two vehicles are usually complicated events and the determination of pre-crash scenarios from pre-crash variables (*Movement Prior to Critical Event* and *Critical Event*) is difficult.

2.2. Rear-End Crashes and Pre-Crash Scenarios

In the simplest model of a rear-end crash, the front of a following vehicle strikes the rear of a lead vehicle, both traveling in the same direction. More complicated cases could involve three or more vehicles piling up. There were 1,806,000 PR rear-end crashes involving light vehicles in 2000 based on GES statistics. Of these, 1,513,000 crashes, or 84 percent, involved 2 vehicles. The remaining 293,000 crashes, or 16 percent, involved 3 or more vehicles.

Six scenarios were identified for rear-end crashes as listed in Table 2-3 and illustrated in Figure 2-2, based on the pre-crash movements of either the lead vehicle or the following vehicle. In 59.1 percent of all rear-end crashes or 895,000 crashes, the lead vehicle was stopped (whether at a red light, at a stop sign, in a turn lane, attempting to pull into a parking position, stopped due to traffic congestion, or broken down). The significance of this result is that these are not cases of the driver of the following vehicle failing to judge the speed of the lead vehicle. It could be that the driver of the following vehicle was inattentive and did not see the stopped lead vehicle in time to stop, or the lead vehicle was not visible, or a number of other possibilities.

Table 2-3. Distribution of Pre-Crash Scenarios for Rear-End Crashes of Light Vehicles

Lead Vehicle Changing Lanes	Following Vehicle Changing Lanes	Lead Vehicle Decelerating	Lead Vehicle Accelerating	Lead Vehicle Stopped	Lead Vehicle Moving at Constant Speed	Total, all scenarios
25,000	30,000	401,000	17,000	895,000	144,000	1,513,000
1.6%	2.0%	26.5%	1.1%	59.1%	9.5%	100.0%

The next most common scenario, accounting for 26.5 percent of all rear-end crashes (a total of 401,000 crashes), involves a lead vehicle decelerating. Once again, the reason for deceleration could be anything from stopping at a red light, making a turn, pulling into a driveway or parallel parking location, to any other possible reason for slowing down. In these cases, driver inattention, following too closely, or misjudgment of the speed of the lead vehicle by the driver of the following vehicle are possible factors in the crash.

There is a certain amount of overlap between these two top scenarios, in the sense that lead vehicle stopped crashes might involve a vehicle that decelerates to a stop immediately before the crash. In fact, the lead vehicle has just decelerated and stopped to either make a turn or comply with a traffic control device in about 51.7 percent of the lead vehicle stopped crashes or 463,000 rear-end crashes. If 401,000 lead vehicle decelerating crashes are added, then the population of the lead vehicle decelerating prior to impact amounts to 864,000, or 57.1 percent of all 2-vehicle rear-end crashes.

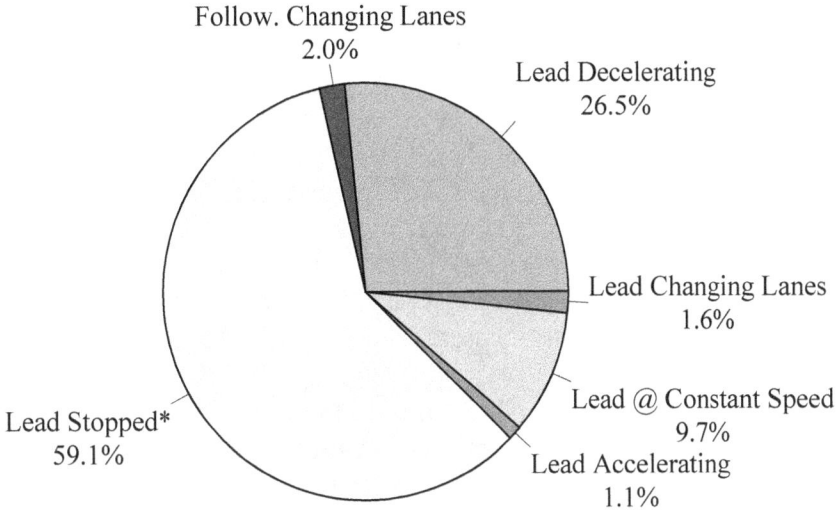

*: Lead vehicle just decelerated and stopped in 51.7% of lead vehicle stopped crashes.

Figure 2-2. Distribution of Pre-Crash Scenarios for Rear-End Crashes of Light Vehicles

The only other significant scenario consists of a lead vehicle moving at constant speed. The following vehicle could either be moving at a higher constant speed, or accelerating. The remaining three scenarios, while relatively few in number, are interesting in terms of the vehicle movements involved. Two of them involve lane change maneuvers: a faster vehicle changing into the lane of a slower vehicle and colliding with it from the rear (2 percent of rear-end crashes), or a slower vehicle changing into the lane of a faster vehicle and being struck in the rear (1.6 percent of rear-end crashes). In the smallest scenario (1.1 percent of rear-end crashes), the lead vehicle is actually accelerating, which means that the following vehicle was either moving at a greater constant speed, or accelerating at a greater rate. This could happen, conceivably, when two vehicles are stopped for a red light, and the lead vehicle accelerates less than the following vehicle when the light changes. (This is just one hypothetical example, intended merely to illustrate not to exactly define the scenario.)

2.3. Crossing Path Crashes and Pre-Crash Scenarios

In a crossing path crash, one moving vehicle cuts across the path of another, initially approaching from either lateral or opposite directions, in such a way that they collide at or near a junction. The number of crossing path crashes in 2000 amounted to 1,590,000 or about 26 percent of all light vehicle crashes. The specific ways in which a crossing path crash might occur are evident from the titles of the pre-crash scenarios shown in Table 2-4, Figure 2-3, and Figure 2-4. Note that the numbers shown reflect all crashes regardless of the number of vehicles involved. The GES *Accident Type* variable, instead of the pre-crash variables, was utilized to determine the various crossing path pre-crash scenarios. Thus, crashes involving more than two vehicles per crash could be included. The following illustrative descriptions and examples of the scenarios are restricted to two-vehicle crashes for simplicity. Figure 2-4 illustrates the scenarios, again using two vehicles for simplicity, on a four-way, perpendicular intersection – this is not necessarily the case in the GES data, of course.

Table 2-4 and Figure 2-3 show a distribution of six known pre-crash scenarios common in crossing path light vehicle crashes, ranked below in a descending order:

1. Straight Crossing Paths (SCP): 545,000 crashes
2. Left Turn Across Path – Opposite Direction (LTAP/OD): 425,000 crashes
3. Left Turn Across Path – Lateral Direction (LTAP/LD): 306,000 crashes
4. Left Turn Into Path – Merge (LTIP): 94,000 crashes
5. Right Turn Into Path – Merge (RTIP): 93,000 crashes
6. Right Turn Across Path – Lateral Direction (RTAP/LD): 34,000 crashes

Table 2-4. Distribution of Pre-Crash Scenarios for Crossing Path Crashes of Light Vehicles (All light vehicle crashes, no restrictions on number of vehicles involved, based on GES 2000)

Left Turn Across Path/ Opposite Direction	Left Turn Into Path	Right Turn Into Path	Right Turn Across Path	Left Turn Across Path/ Lateral Direction	Straight Crossing Paths	Other/ Unknown	Total, All Scenarios
425,000	94,000	93,000	34,000	306,000	545,000	93,000	1,590,000
26.8%	5.9%	5.8%	2.1%	19.2%	34.3%	5.9%	100.0%

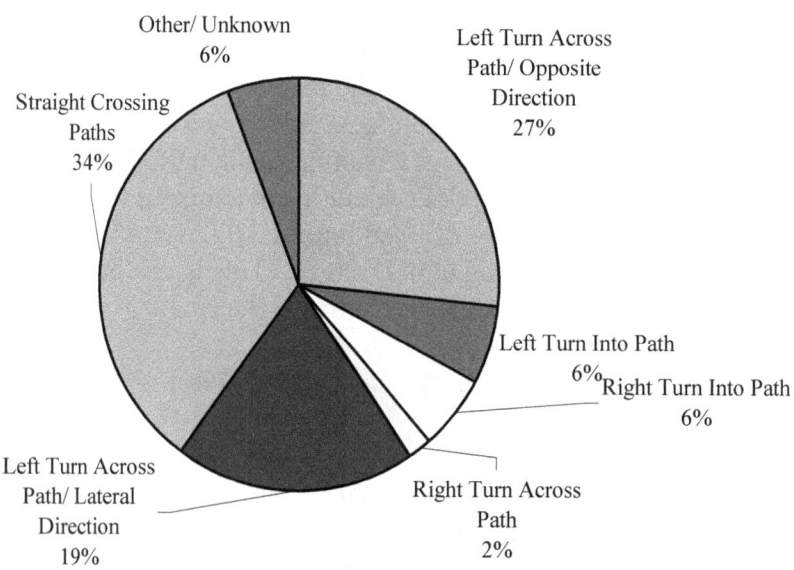

Figure 2-3. Distribution of Pre-Crash Scenarios for Crossing Path Crashes of Light Vehicles

The SCP scenario dominates all crossing path crashes at about 34 percent of this crash type. This scenario involves vehicles going straight on lateral intersecting paths, including for instance, two vehicles that approach an unsignalized intersection, both intending to go straight, and neither of the drivers yields to the other. Another illustrative example of this crash type is two vehicles approaching a signalized intersection from lateral directions, with both drivers intending to go straight, and one driver running a red light.

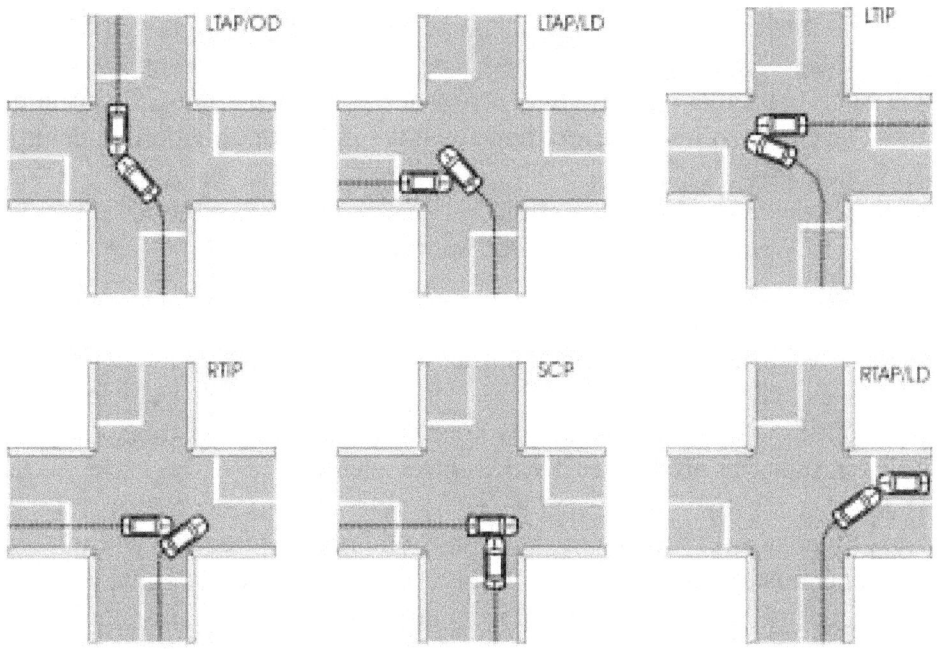

Figure 2-4. Schematics of Common Crossing Path Pre-Crash Scenarios

The LTAP/OD scenario is the next most common scenario, which constitutes about 27 percent of all light vehicle crossing path crashes. This scenario consists of vehicles on initial opposite directions, with one or more of them turning left across the path of the others. An example is a crash occurring at an intersection without a protected left turn signal, in which two vehicles approach the intersection from opposite directions, one intending to go straight and one turning left, where the driver turning left does not yield to the driver going straight (or misjudges their speed), leading to a collision.

The third most common scenario is the LTAP/LD at about 19 percent of the crash type, which consists of vehicles on initial lateral directions, with one or more of them turning left across the path of the other. To clarify, the vehicle or vehicles going straight are approaching the left-turning vehicle or vehicles *from the left* as shown in Figure 2-4. Some ways in which this scenario could occur include signalized intersections in which one of the vehicles (either the turning vehicle or the vehicle going straight) is running a red light, or unsignalized intersections with one vehicle failing to yield right of way, or a vehicle pulling out of a driveway and turning left, colliding with a vehicle traveling in the lane across which it is turning.

The next two scenarios in size, both accounting for about 6 percent of crossing path crashes, are the LTIP/LD and RTIP/LD crashes, in which two vehicles are traveling in initial lateral directions, and one turns left or right into the same direction as the other. Some common situations could be a vehicle turning right on red (either at an intersection where it is allowed, or in violation of a "No Turn On Red" sign); a vehicle making a left turn at a T intersection; an unsignalized intersection; the driver going straight running a red light; the turning vehicle pulling out of a driveway; etc.

The smallest known scenario is the RTAP/LD scenario. A particularly interesting feature of this scenario is that it involves a driving maneuver error other than only a signal violation, a failure to yield right of way, a misjudgment of speed of another vehicle, or a vision/attention problem (failure to see another vehicle on a crossing path), all of which are common features in all of the other scenarios. This particular scenario also involves *encroachment*; i.e., the turning vehicle fails to stay in its lane and encroaches into the lane in the opposite direction while turning, as clearly seen in Figure 2-4.

2.4. Off-Roadway Crashes and Pre-Crash Scenarios

Off-roadway crashes refer to crashes in which the first harmful event occurs off the roadway after a vehicle in transport departs the travel portion of the roadway. These crashes totaled 1,280,000 PR crashes in 2000, comprising almost 21 percent of all light vehicle crashes. Unlike other crash types discussed so far, off-roadway crashes are divided into crash subtypes in this study, and the crash subtypes are further divided into scenarios.

Table 2-5(a) and Figure 2-5(a) provide the frequency and relative frequency of off-roadway crash subtypes. The most dominant subtype involves a single vehicle running off the roadway while going forward, and colliding with an object and/or overturning. This single crash subtype accounts for 88 percent of all light vehicle off-roadway crashes. About 6.3 percent of off-roadway crashes consist of multi-vehicle collisions, while 5.5 percent involve a vehicle backing into another vehicle or object not on the roadway. Only 0.3 percent of off-roadway crashes are of the no impact subtype, which includes immersion, fires, etc.

The discussion of pre-crash scenarios is confined to the single vehicle crash subtype, since it encompasses the vast majority of off-roadway crashes. The pre-crash scenarios were defined using both the *Movement Prior to Critical Event* and *Critical Event* variables. There is an implicit matrix of these two variables in the scenario classification, for at least eight of the scenarios. The *Movement Prior to Critical Event* variable could indicate that the vehicle was going straight, or that it was negotiating a curve. It is useful to distinguish these, because maintaining control (and therefore staying on the roadway), or following the lane properly, is much harder on a curve than on a straight road, especially at higher speeds. This differentiates off-roadway crashes while going straight, from off-roadway crashes while negotiating a curve. Alternatively, the *Movement Prior to Critical Event* variable could indicate that the vehicle was not simply going forward either straight or on a curve, but was initiating a special driving maneuver such as a lane change, passing, turning, or merging movement. All of these movements are at a complexity and difficulty level greater even than negotiating a curve. Finally, there are cases in which the vehicle movement was not any of the above three; namely, going straight, negotiating a curve, or initiating a maneuver.

Similarly, the *Critical Event* variable could indicate two distinct possibilities. A vehicle could be coded as losing control due to either speeding or poor road conditions, leading to departure from the roadway. Alternatively, a vehicle could be in motion, and deviate from the lane and depart the road edge even without any catastrophic loss of control; this constitutes road edge departure in our scenario definition.

Table 2-5(b) and Figure 2-5(b) define 10 single vehicle off-roadway pre-crash scenarios. Eight of these scenarios resulted from all possible combinations of the movement prior to critical event and critical event categories discussed above. The remaining two scenarios involve vehicle failure (i.e., catastrophic loss of control leading to road edge departure, caused by vehicle component failure such as brake failure or tire blowout) and avoidance maneuvers in which a driver attempts to avoid another type of crash, such as a rear-end collision, and runs off the roadway.

In as many as one out of four single vehicle off-roadway crashes, a vehicle was going straight and simply departed the road edge for no apparent reason. There was no vehicle failure, avoidance maneuver, driving maneuver, curve in the roadway, or loss of control. It is not unreasonable to speculate that alcohol, drugs, fatigue/drowsiness, or inattention had played a major role in a large number of these crashes.

Table 2-5(a). Distribution of Crash Subtypes for Off-Roadway Crashes of Light Vehicles (Based on GES 2000)

Single Vehicle	Backing	No Impact	Multi-Vehicle	Total
1,126,000	70,000	3,000	81,000	1,280,000
88.0%	5.5%	0.3%	6.3%	100.0%

Note: This table, unlike the other tables in this section, shows the distribution of crash subtypes rather than pre-crash scenarios. A crash subtype is based on number of vehicles involved and critical event, rather than pre-crash sequence of events as described by the *Movement Prior to Critical Event* variable. Each crash subtype for off-roadway crashes can be further divided into pre-crash scenarios.

Table 2-5(b). Distribution of Pre-Crash Scenarios for Single Vehicle Off-Roadway Crashes of Light Vehicles (Based on GES 2000)

Vehicle Failure	Going Straight and Lost Control	Negotiating a Curve and Lost Control	Initiating a Maneuver and Lost Control	Other Control Loss	Going Straight and Departed Road Edge	Negotiating a Curve and Departed Road Edge	Initiating a Maneuver and Departed Road Edge	Other Road Edge Departure	Avoidance Maneuvers	Other	Total
32,000	220,000	165,000	48,000	3,000	281,000	110,000	54,000	8,000	118,000	87,000	1,126,000
2.9%	19.6%	14.6%	4.3%	0.3%	25.0%	9.8%	4.8%	0.7%	10.5%	7.7%	100.0%

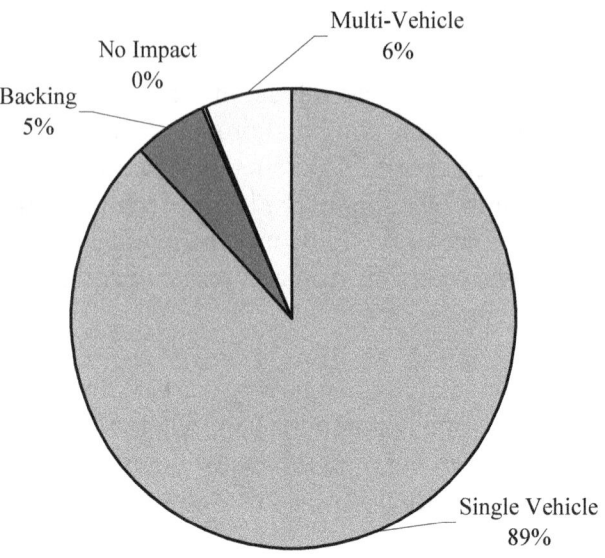

Figure 2-5(a). Distribution of Crash Subtypes for Off-Roadway Crashes of Light Vehicles

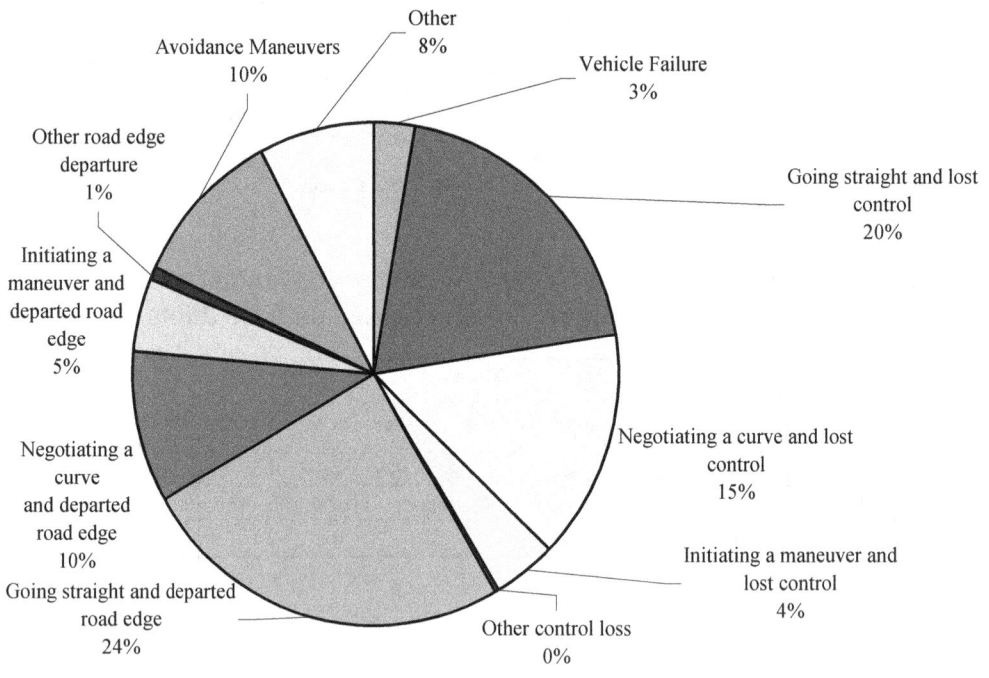

Figure 2-5(b). Distribution of Pre-Crash Scenarios for Single Vehicle Off-Roadway Crashes of Light Vehicles

A vehicle was going straight and lost control in about 20 percent of all single light vehicle off-roadway crashes. This second largest scenario can be attributed to many factors including speeding, and poor roadway surface conditions.

The third and fourth largest scenarios involve a vehicle negotiating a curve and then losing control or simply departing the edge of the road in about 15 percent and 10 percent of all single light vehicle off-roadway crashes, respectively. The relative dominance of single vehicle off-roadway crashes with the vehicle going straight, as opposed to negotiating a curve, probably has more to do with relative exposure than anything else. It is far more common for vehicles to be traveling straight rather than negotiating a curve. However, it is quite possible that the likelihood of a vehicle going off the roadway while negotiating a curve is disproportionately high; i.e., the share of off-roadway crashes involving a vehicle negotiating a curve is much higher than the share of driving that occurs on curved roadways. This is because, as stated earlier, negotiating a curve is a more complex operation than going straight, with greater likelihood of losing control at higher speeds or being unable to keep the vehicle in lane and going off the roadway.

A little less than 11 percent of single light vehicle off-roadway crashes involve avoidance maneuvers. The 118,000 avoidance maneuver off-roadway crashes consisted of maneuvers to avoid 39,000 animal/object collisions, 29,000 lane change crashes, 28,000 opposite direction crashes, 18,000 rear-end crashes, and 4,000 pedestrian/pedalcyclist crashes. For the remaining 87,000 avoidance maneuver off-roadway crashes, the reason for the avoidance maneuver was unknown.

2.5. Lane Change Crashes and Pre-Crash Scenarios

The lane change family of crashes typically consists of a crash in which a vehicle attempts to change lanes, merge, pass, leave/enter a parking position, drifts and strikes, or is struck by another vehicle in the adjacent lane, both traveling in the same direction. The condition of initial travel in the same direction is the key distinguishing feature of lane change crashes from crossing paths and opposite direction crashes. In 2000, there were 565,000 lane change crashes involving light vehicles (9.2 percent of all light vehicle crashes). Of these, 545,000 crashes (96.5 percent) involved 2 vehicles.

There are many possible combinations of vehicle movements and critical events that could lead to a lane change crash, as shown in Table 2-6(a). This matrix shows the distribution of lane change crashes by different combinations of vehicle movements, and different critical events. Each cell in the matrix represents a pre-crash scenario. Table 2-6(b) and Figure 2-6 present the shares of the seven largest scenarios.

Table 2-6(a). Pre-Crash Scenario Matrix for Lane Change Crashes of Light Vehicles (Two-vehicle crashes only; based on GES 2000)

Movement Prior to Critical Event	Critical Event							Total
	E1	E2	E3	E4	E5	E6	E7	
M1 and M1	2,000	19,000	21,000	42,000	0	2,000	16,000	**102,000**
M1 and M2	0	3,000	3,000	22,000	0	0	4,000	**32,000**
M1 and M3	0	0	0	20,000	0	0	6,000	**26,000**
M1 and M4	0	1,000	5,000	76,000	0	0	10,000	**92,000**
M1 and M5	0	3,000	6,000	190,000	0	0	4,000	**203,000**
M1 and M6	0	1,000	0	14,000	0	0	2,000	**17,000**
M1 and M7	0	1,000	2,000	5,000	0	0	4,000	**12,000**
M2 and M4	0	0	3,000	32,000	0	0	9,000	**44,000**
Other (including both vehicles changing lanes, or turning, etc.)	0	1,000	0	9,000	0	0	7,000	**17,000**
Total	**2,000**	**29,000**	**41,000**	**409,000**	**0**	**2,000**	**62,000**	**545,000**

Key for Critical Event: The Respective Columns Include Crashes in Which At Least One of the Vehicles Had The Code Specified.
E1: Vehicle Failure
E2: Other Loss of Control
E3: Another Vehicle in Same Lane; Traveling Faster, Slower, Accelerating, Decelerating, Stopped, etc.
E4: One Vehicle Encroaching Into Another Lane; Another Vehicle Has Critical Event of Other Vehicle Encroaching Into Its Lane
E5: Pedestrian or Pedalcyclist
E6: Animal or Object
E7: Other Codes

Key for Movement Prior to Critical Event: In Each Row, The Two Vehicles Involved Have the Two Respective Codes Specified.
M1: Going Straight
M2: Passing
M3: Entering/Leaving Parked Position
M4: Turning
M5: Simple Lane Change
M6: Merging
M7: Other Movement

Table 2-6(b). Distribution of Major Pre-Crash Scenarios for Lane Change Crashes of Light Vehicles (Two-vehicle crashes only; based on GES 2000)

Scenario Number	1	2	3	4	5	6	7	Other	
Critical Event	*E4*	*E4*	*E4*	*E4*	*E4*	*E3*	*E4*	*All Other Combinations*	**Total, All Scenarios**
Movement Prior to Critical Event	*M1 and M5*	*M1 and M4*	*M1 and M1*	*M2 and M4*	*M1 and M2*	*M1 and M1*	*M1 and M3*		
Number	190,000	76,000	42,000	32,000	22,000	21,000	20,000	142,000	**545,000**
% Share	34.9%	13.9%	7.7%	5.9%	4.0%	3.9%	3.7%	26.0%	**100.0%**

Note: The "All Other Combinations" column should not be interpreted as "other and unknown." It includes all cells from the matrix in Table 2-6(a) that have values below 20,000.

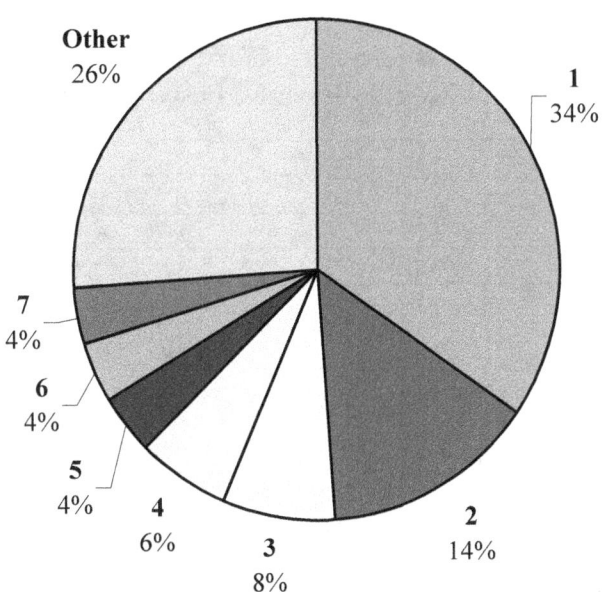

Figure 2-6. Distribution of Major Pre-Crash Scenarios for Lane Change Crashes of Light Vehicles

Six of the seven largest scenarios were characterized by the critical event of one vehicle encroaching into the lane of another vehicle. The encroaching vehicle was performing a simple lane change maneuver in the most common scenario, accounting for almost 35 percent of lane change crashes. In the next highest scenario, the encroaching vehicle was making a turning maneuver at a relative frequency of 14 percent. The third largest scenario does not involve any planned maneuver on the part of the driver of the encroaching vehicle; rather, it appears that the driver failed to stay within the lane and drifted into the adjacent lane. The circumstances could include drugs and alcohol, fatigue/drowsiness, inattention, or failure to keep one's lane (especially at high speeds, on a curve, or both). The three other encroachment scenarios shown in Table 2-6(b) and Figure 2-6, in order of size, are passing combined with turning maneuvers (5.9 percent), passing maneuvers (4.0 percent), and parking maneuvers (3.7 percent). The sixth largest scenario, falling in between the passing/encroachment scenario and the parking/ encroachment scenario (with a share of 3.9 percent), consists of two vehicles going straight, *in the same lane*, with the following vehicle traveling at a higher steady speed than the lead vehicle, or accelerating, or the lead vehicle decelerating or stopped. It is not clear why these crashes are classified as belonging to the lane change accident type (Accident Type codes = 44-49 or 70-73) in the GES.

2.6. Animal Crashes and Pre-Crash Scenarios

The next largest crash type after lane change crashes is crashes with animals, with 247,000 or 4 percent of all light vehicle crashes reported in 2000. As seen in Table 2-7 and Figure 2-7, about

94 percent of these crashes involve a vehicle going straight. The vehicle was negotiating a curve in the next largest scenario that accounted for about 4 percent of light vehicle crashes with animals. The two scenarios are distinguished by roadway alignment because there may be visibility issues in cases with the vehicle negotiating a curve. Due to the curve in the road, the driver of the vehicle may not see the animal until it is too late to stop.

Table 2-7. Distribution of Pre-Crash Scenarios for Crashes of Light Vehicles with Animals (Single vehicle crashes only; based on 2000 GES)

Vehicle Going Straight - Animal in Roadway	Vehicle Negotiating a Curve - Animal in Roadway	Other	Total, all Scenarios
232,000	9,000	6,000	247,000
93.8%	3.8%	2.3%	100.0%

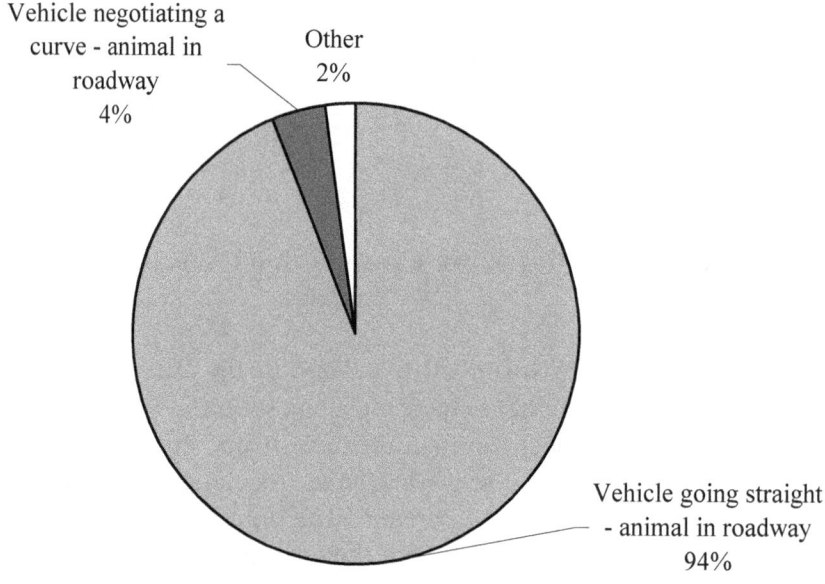

Figure 2-7. Distribution of Pre-Crash Scenarios for Crashes of Light Vehicles with Animals

The remaining scenario, other and unknown, includes cases with the vehicle going straight and the animal approaching the roadway; cases with the vehicle making a maneuver and an animal in the roadway; and other movements that are not classified.

2.7. Opposite Direction Crashes and Pre-Crash Scenarios

Opposite direction crashes involve a vehicle that strikes another vehicle in the adjacent lane, traveling in the opposite direction. These crashes result in a frontal or sideswipe impact. This crash type could also conceivably involve vehicles traveling the wrong way on a one-way street.

Light vehicles were associated with 163,000 PR opposite direction crashes based on 2000 GES estimates, accounting for 2.7 percent of all PR light vehicle crashes.

As with single vehicle off-roadway crashes, a distinction was made between vehicles negotiating a curve and vehicles going straight for the purpose of identifying pre-crash scenarios. The reasons for such a distinction are as follows:

- On a curved roadway, a vehicle in the same lane going in the opposite direction may not be visible until it is too late to stop.
- Vehicles are harder to control on a curve, especially at higher speeds, than on a straight roadway.
- The likelihood of failing to keep to one's lane and encroaching across the median line is greater on a curved road, with high speeds as a risk factor.

Another distinction was adopted to separate the cases in which a vehicle encroached into the lane from the opposite direction shortly before the crash from the cases in which a vehicle had apparently been in the wrong lane for a while before the crash. The latter may include cases of drivers going the wrong way on a one-way street, and may also include a significant share of cases of alcohol impairment.

Table 2-8 and Figure 2-8 provide the results of opposite direction pre-crash scenarios. The share of opposite direction crashes in which the vehicles were negotiating a curve corresponds to at least 38 percent of all opposite direction crashes. It should be noted that the relative frequency of opposite direction crashes on curves might actually be greater than 38 percent because our analysis did not differentiate control loss, vehicle failure, passing, and other and unknown crashes by the roadway alignment. Cases of encroachment shortly prior to a crash far outweigh cases in which one vehicle was in the wrong lane for some time before the crash, for vehicles going straight or negotiating a curve. About 82 percent of opposite direction crashes involve a vehicle encroaching into the wrong lane shortly before the crash. Counterintuitively, the share of the passing scenario in opposite direction crashes is actually quite small (3 percent).

Table 2-8. Distribution of Pre-Crash Scenarios for Opposite Direction Crashes of Light Vehicles (Two-vehicle crashes only; based on GES 2000)

Going Straight/ Encroaching	Going Straight/ In Lane	Negotiating a Curve/Encroaching	Negotiating a Curve/In Lane	Control Loss (Except when Passing)	Passing	Vehicle Failure	Other	Total, All Scenarios
74,000	7,000	60,000	3,000	1,000	5,000	1,000	12,000	**163,000**
45.4%	4.6%	36.4%	1.8%	0.7%	3.1%	0.9%	7.2%	**100.0%**

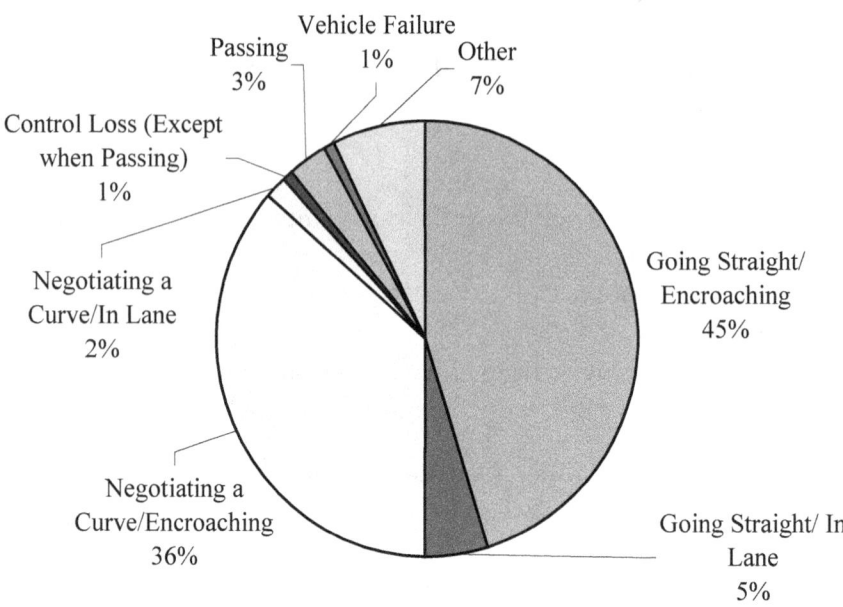

Figure 2-8. Distribution of Pre-Crash Scenarios for Opposite Direction Crashes of Light Vehicles

2.8. Backing Crashes and Pre-Crash Scenarios

A backing crash is defined as a vehicle striking or struck by an obstacle or another vehicle while moving backwards. The GES classifies backing crashes as a unique crash type. The definition excludes crashes with pedestrians and pedalcyclists. In 2000, there were 127,000 PR backing crashes, or 2.1 percent of all PR light vehicle crashes, based on GES estimates.

The manner in which pre-crash scenarios have been identified for backing crashes implicitly accounts for the relation to junction, as indicated in Table 2-9 and Figure 2-9. The *Movement Prior to Critical Event* and *Critical Event* variables do not contain sufficient information to determine whether a vehicle was backing in a straight line or was backing while turning. Crashes involving a vehicle backing from a driveway were considered unique because of the associated constraints on driver field of vision, and because the backing vehicle and the other vehicle were not initially on the same roadway. Crashes at intersections were also considered unique, because they involve *at least some* cases in which the driver of the backing vehicle cannot see the other vehicle, and/or in which the backing vehicle and the other vehicle were not initially on the same roadway. Driveway backing crashes, and some intersection backing crashes, would fall into the "Curved Path" and "Straight Crossing Paths" categories as defined in Reference [15]. The remainder, or other/unknown backing crashes, would mostly fall into the "Parallel Paths" category [15].

**Table 2-9. Distribution of Pre-Crash Scenarios for Backing Crashes of Light Vehicles
(Two-vehicle crashes only; based on 2000 GES)**

Intersection Crashes	Driveway Crashes	Other Crashes	Total Crashes
47,000	50,000	30,000	**127,000**
37.0%	39.1%	23.9%	**100.0%**

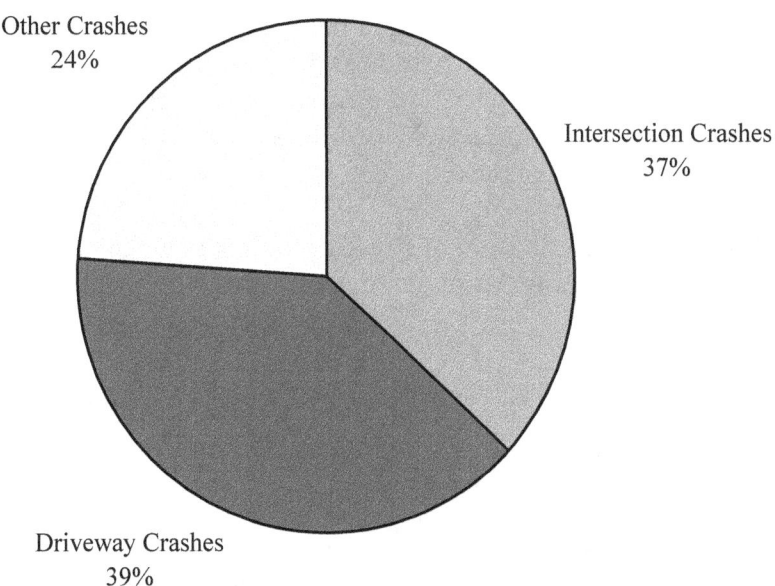

Figure 2-9. Distribution of Pre-Crash Scenarios for Backing Crashes of Light Vehicles

The 2000 GES did not report any backing crash cases in which the vehicle lost control. This might be attributed to backing maneuvers that typically occur at low speeds, thus reducing the probability of a catastrophic loss of control.

2.9. Pedestrian Crashes and Pre-Crash Scenarios

A pedestrian crash involves a moving vehicle colliding with a pedestrian. As with animal crashes, discussed in Section 2.6, this analysis considers pedestrian crashes as *single vehicle* crashes. The definition of pedestrian crashes excludes cases in which a vehicle tries to avoid a pedestrian and, as a result of the avoidance maneuver, collides with another vehicle or an object. Based on GES statistics, a total of 66,000, or 1.1 percent of all PR light vehicle crashes, were reported as pedestrian crashes in 2000.

The breakdown of pedestrian crashes into pre-crash scenarios was accomplished by unique combinations of vehicle maneuvers and pedestrian movements. As seen in Table 2-10 and Figure 2-10, vehicle maneuvers include going straight, turning right, turning left, and backing.

Pedestrian movements include a pedestrian crossing the roadway, darting onto the roadway, walking along the roadway, not in the roadway, and playing or working in the roadway. Eight common pre-crash scenarios were identified with a combined relative frequency of about 91 percent. The most common scenario involves a vehicle going straight and a pedestrian crossing the roadway, which accounted for about 38 percent of all light vehicle crashes with pedestrians. The second leading scenario associates a vehicle going straight with a pedestrian darting onto the road. The GES distinguishes between pedestrian crossing the roadway and pedestrian darting onto the roadway. A dart-out is coded if it is documented that the driver's view of the pedestrian was obstructed by a physical object such as a bus, stopped or parked vehicle, or a building.

A vehicle was going straight in about 73 percent of all light vehicle crashes with pedestrians. A vehicle was turning in about 20 percent of these crashes. It should be noted that Scenario 9 in Table 2-10 (other) includes some cases of vehicles going straight or turning in connection with other pedestrian movements and actions. A pedestrian was crossing the roadway in about 60 percent of all light vehicle crashes with pedestrians. A pedestrian darting onto the roadway was associated with 23 percent of these crashes.

Table 2-10. Distribution of Pre-Crash Scenarios for Crashes of Light Vehicles with Pedestrians (Single-vehicle crashes only; based on 2000 GES)

No.	Scenario	Frequency	
1	Vehicle going straight and pedestrian crossing the road	25,000	37.9%
2	Vehicle going straight and pedestrian darting onto the road	15,000	22.7%
3	Vehicle turning left and pedestrian crossing the road	7,000	10.6%
4	Vehicle turning right and pedestrian crossing the road	4,000	6.1%
5	Vehicle is going straight and pedestrian is doing "other"	3,000	4.5%
6	Vehicle going straight and pedestrian walking along the road	2,000	3.1%
7	Vehicle going straight and pedestrian playing/working in the road	2,000	3.0%
8	Vehicle backing	2,000	3.0%
9	Other	6,000	9.1%
Total		66,000	100.0%

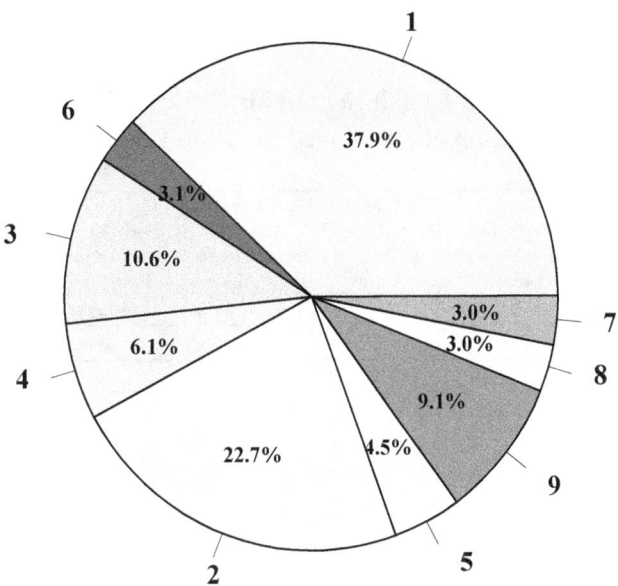

Figure 2-10. Distribution of Pre-Crash Scenarios for Crashes of Light Vehicles with Pedestrians

2.10. Pedalcyclist Crashes and Pre-Crash Scenarios

A pedalcyclist crash occurs when a motor vehicle strikes or is struck by a pedalcyclist. A collision of a motor vehicle with another motor vehicle or object, arising from an avoidance maneuver in which the driver tries to avoid a collision with a pedalcyclist, is not included in this population of pedalcyclist crashes. The 2000 GES reported 47,000 PR pedalcyclist crashes with light vehicles, or 0.8 percent of all light vehicle crashes.

Pedalcyclist crashes were broken down into eight pre-crash scenarios based on combinations of the vehicle's movement prior to the critical event and the initial approach path (parallel, crossing, other) of the pedalcyclist in relation to the vehicle's maneuver. "Parallel paths" were defined as cases where the cycle and motor vehicle were approaching each other on parallel paths, heading either in the same or in opposite directions. "Crossing paths" include cases where the cycle and the motor vehicle were on intersecting paths. Cases were classified as "other/unknown" when it was unknown whether the vehicle and cycle's initial approach paths were parallel or crossing, or the crash involved a vehicle performing a other maneuver (i.e., backing, parking, passing, changing lanes, or an avoidance maneuver). The eight pre-crash scenarios fully encompassed the entire pedalcyclist crash type.

Pedalcyclist crashes occur in circumstances similar to a number of crash types and concomitant pre-crash scenarios discussed earlier, especially crossing path crashes. As seen in Table 2-11 and Figure 2-11, the most common scenario for this crash type was when the vehicle was traveling straight on a crossing path with the pedalcyclist, accounting for over two-fifths of all pedalcyclist crashes with light vehicles. This scenario dynamically resembles the corresponding scenario (SCP) for crossing path crashes. The circumstances may also be similar: one driver or the cyclist may be running a red light or a stop sign, or one or both of them may have failed to see the other approaching the intersection.

Table 2-11. Distribution of Pre-Crash Scenarios for Crashes of Light Vehicles with Pedalcyclists (based on 2000 GES)

Vehicle Traveling Straight/Crossing Paths	Vehicle Traveling Straight/Parallel Paths	Vehicle Turning Right/Crossing Paths	Vehicle Turning Right/Parallel Paths	Vehicle Turning Left/Crossing Paths	Vehicle Turning Left/Paralle Paths	Vehicle Starting in Traffic/Crossing Paths	Other/ Unknown	Total
21,000	7,000	5,000	3,000	3,000	2,000	3,000	3,000	**47,000**
43.6%	14.6%	11.3%	5.7%	6.5%	4.5%	6.5%	7.3%	**100.0%**

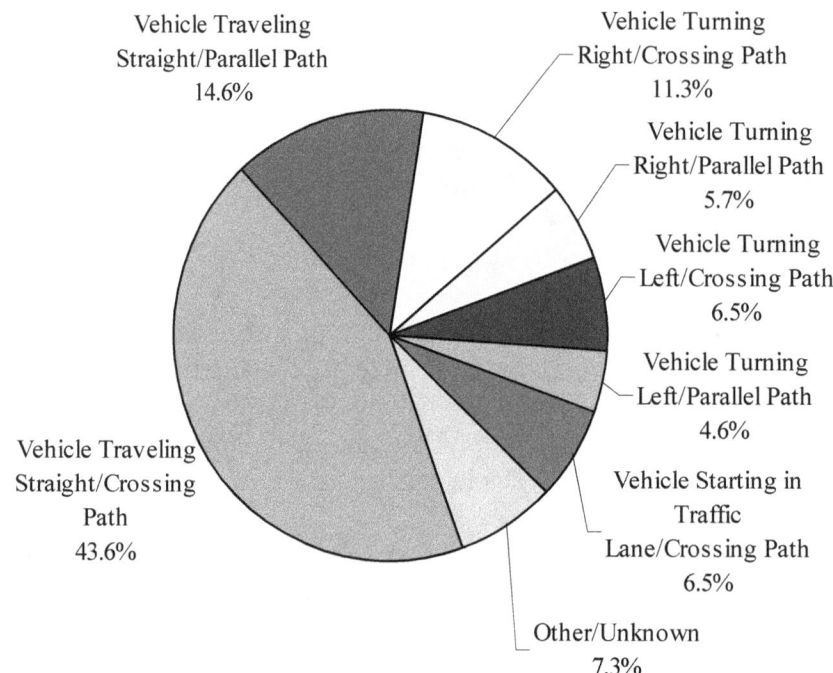

Figure 2-11. Distribution of Pre-Crash Scenarios for Crashes of Light Vehicles with Pedalcyclists

Other common pedalcyclist pre-crash scenarios include cases where the cyclist approaches the vehicle on a parallel path in which the vehicle is traveling straight. Crashes involving the vehicle turning right are more prevalent when the vehicle and the cyclist are on crossing paths than on parallel paths, 11.3% compared to 5.7%. The same was true for crashes involving the vehicle turning left, 6.5% of the crashes occurred when the cyclist was on a crossing path compared to 4.6% while on a parallel path. Overall, 68% or the pedalcyclist crashes occurred on crossing paths and 25% occurred on parallel paths.

3. PHYSICAL SETTING OF CRASHES AND PRE-CRASH SCENARIOS

This section describes the distribution of the major crash types discussed in Section 2 with respect to the *Relation to Junction* variable from the GES. It also describes the distribution with respect to relation to junction of selected pre-crash scenarios for each of the crash types. The selection of pre-crash scenarios for discussion of their distribution by relation to junction is based both on the relative size of the scenario, and by the degree of their difference with each other and with the overall distribution for the crash type. For example, the top two scenarios for lane change crashes are selected for discussion of their distribution with respect to relation to junction, both because they are the two largest scenarios, and because their respective distributions are significantly different from each other as well as from the overall distribution for lane change crash types (see Table 3-5, and Figures 3.5(a), (b), and (c)).

The *Relation to Junction* variable in the GES indicates whether or not the location of the first harmful event occurred at different types of junctions including intersection, intersection-related, driveway or alley access, entrance or exit ramp, rail grade crossing, on a bridge, and non-junction. The GES defines an intersection as that area enclosed by the extension of the lateral curb lines of the intersecting roadways. For a crash to be coded as intersection-related, the first harmful event must occur on the segments of roads leading to the intersection and must be related to motion through the intersection. The codes for "driveway or alley access," "ramp," and "grade crossing" mean that the crash must be related to motion through a junction between these and a roadway.

The distribution of crash types by relation to junction for crashes of all vehicle types is shown in Table 3-1(a) and Figure 3-1(a). The corresponding distribution for crashes involving light vehicles is shown in Table 3-1(b) and Figure 3-1(b). The differences in the distributions for crashes of all vehicle types and light vehicle crashes are negligible. About 45 percent of all crashes are seen to occur at or near intersections, and another 40 percent at non-junction locations. The next most common location is driveways, which account for about 11 percent of all crashes. Intersection, intersection-related, and driveway locations combined account for about 55 percent of all crashes. These percentages are applicable to all-vehicle and light-vehicle crashes.

Table 3-1(a). Distribution of All Vehicle Crashes by Relation to Junction (based on GES 2000)

Relation to Junction	Number of Crashes	Share by Relation to Junction
Non-Junction	2,595,000	40.6%
Intersection	1,518,000	23.8%
Intersection-Related	1,289,000	20.2%
Driveway/Alley	677,000	10.6%
Entrance/Exit Ramp	164,000	2.6%
Rail Grade Crossing	15,000	0.2%
On a Bridge	57,000	0.9%
Crossover Related	19,000	0.3%
Other	55,000	0.9%
Total	**6,389,000**	**100.0%**

Table 3-1(b). Distribution of Light Vehicle Crashes by Relation to Junction (based on GES 2000)

Relation to Junction	Number of Crashes	Share by Relation to Junction
Non-Junction	2,426,000	39.6%
Intersection	1,503,000	24.5%
Intersection-Rel.	1,249,000	20.4%
Driveway/Alley	663,000	10.8%
Ramp	156,000	2.5%
Grade Crossing	14,000	0.2%
On a Bridge	53,000	0.9%
Crossover Rel.	19,000	0.3%
Other	50,000	0.8%
Total	**6,133,000**	**100.0%**

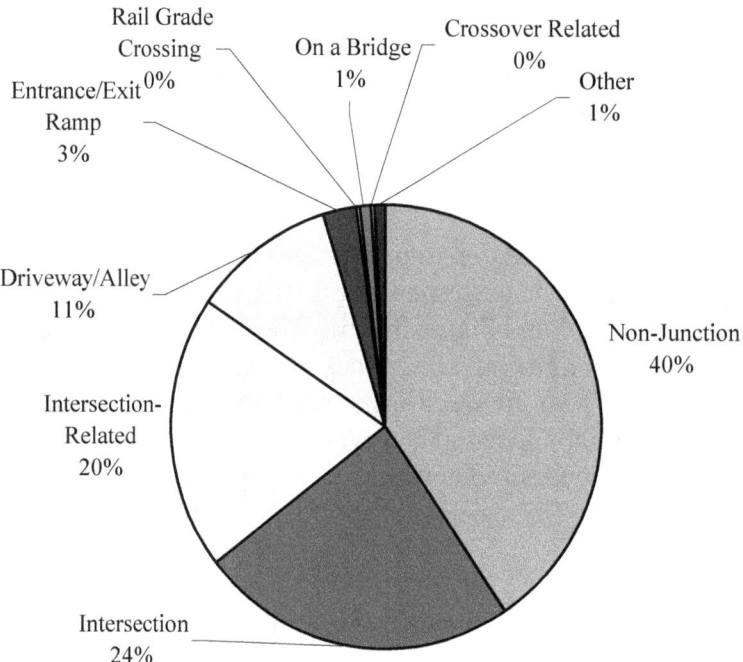

Figure 3-1(a). Distribution of All Vehicle Crashes by Relation to Junction

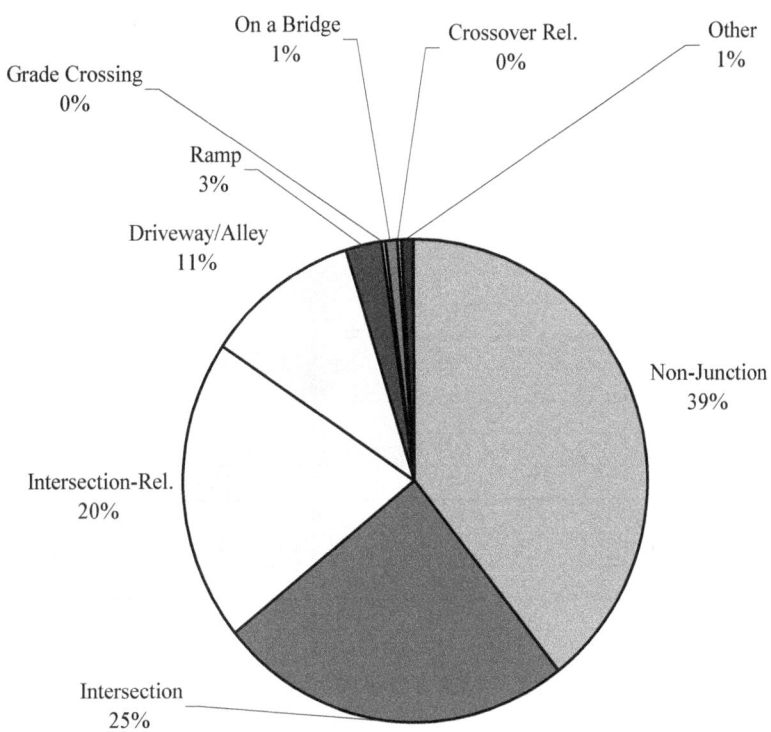

Figure 3-1(b). Distribution of Light Vehicle Crashes by Relation to Junction

This section does not discuss the distribution of backing crashes by relation to junction since this information was implicit in backing pre-crash scenarios as shown in Table 2-9 and Figure 2-8. Finally, this section concludes with some statistics about the traffic control device that was present at crash locations, especially at intersections and intersection-related roadways.

3.1. Rear-End Crashes

As discussed in Section 2, rear-end crashes comprise the largest crash frequency for all vehicles and for light vehicles. About 51.9 percent of two-vehicle rear-end crashes occur at intersections or intersection-related locations as seen in Table 3-2 and Figure 3-2(a). The highest rear-end crash frequency is reported on intersection-related roadways. About 7.1 percent of two-vehicle rear-end crashes happen at driveways/alleys. Thus, a total of 59 percent of rear-end crashes occur at or near junctions of two or more streams of traffic. The significance of this observation becomes clearer when examining the distribution by relation to junction of the various pre-crash scenarios for rear-end crashes (Table 3-2). It should be noted that this study did not analyze the distribution by relation to junction of rear-end crashes involving more than two vehicles.

Table 3-2. Distribution of Pre-Crash Scenarios for Rear-End Crashes of Light Vehicles by Relation to Junction (Two-vehicle crashes only; based on GES 2000)

RELATION TO JUNCTION	Lead Vehicle Changing Lanes	Following Vehicle Changing Lanes	Lead Vehicle Decelerating	Lead Vehicle Accelerating	Lead Vehicle Stopped	Lead Vehicle Moving at Constant Speed	Total, All Scenarios
Non-Junction	14,000	15,000	185,000	3,000	215,000	70,000	**502,000**
Intersection	2,000	1,000	23,000	3,000	60,000	13,000	**101,000**
Intersection-Related	5,000	11,000	122,000	10,000	499,000	36,000	**684,000**
Driveway/Alley	2,000	2,000	40,000	-	48,000	15,000	**107,000**
Entrance/Exit Ramp	1,000	-	19,000	1,000	41,000	6,000	**68,000**
Rail Grade Crossing	-	-	3,000	-	3,000	1,000	**7,000**
On a Bridge	1,000	-	4,000	-	6,000	2,000	**12,000**
Crossover Related	-	-	1,000	-	2,000	1,000	**4,000**
Other	-	-	5,000	-	21,000	2,000	**29,000**
Total by scenario	**25,000**	**30,000**	**401,000**	**17,000**	**895,000**	**144,000**	**1,513,000**
Share by scenario	**1.6%**	**2.0%**	**26.5%**	**1.1%**	**59.1%**	**9.5%**	**100.0%**
Non-Junction	56.5%	50.6%	46.0%	16.2%	24.0%	48.5%	**33.2%**
Intersection	8.9%	2.8%	5.7%	17.2%	6.7%	8.7%	**6.7%**
Intersection-Related	19.8%	36.4%	30.5%	58.1%	55.8%	25.2%	**45.2%**
Driveway/Alley	7.5%	6.6%	10.0%	2.1%	5.4%	10.1%	**7.1%**
Entrance/Exit Ramp	2.5%	1.6%	4.8%	4.1%	4.6%	4.0%	**4.5%**
Rail Grade Crossing	0.0%	0.0%	0.8%	0.4%	0.4%	0.4%	**0.5%**
On a Bridge	3.6%	0.0%	0.9%	0.0%	0.7%	1.1%	**0.8%**
Crossover Related	1.2%	0.5%	0.2%	0.5%	0.2%	0.4%	**0.3%**
Other	0.1%	1.5%	1.1%	1.5%	2.4%	1.6%	**1.9%**

Only the top 2 scenarios are considered for further analysis (Figures 3.2(b) and 3.2(c)); together, they constitute almost 86 percent of all 2-vehicle rear-end crashes of light vehicles and they exhibit sufficiently different distributions with respect to relation to junction. More than 68 percent of lead vehicle stopped rear-end crashes occur at junction locations (55.8 percent intersection related, 6.7 percent at intersections, 5.4 percent at junctions of driveways), while only 24 percent of these crashes occur at non-junction locations as illustrated in Figure 3-2(b). This is consistent with how these crashes are thought to occur; in most of them, the lead vehicle is presumably either stopping/stopped at a red light or stop sign, stopping/stopped to make a turn or to pull into a driveway, or is stopping/stopped due to congestion (which is more likely to happen close to a busy intersection). The crashes occurring at non-junction locations could also include lead vehicles stopped because of congestion, and lead vehicles stopped to pull into roadside parking positions, or stopped in a traffic lane because of vehicle breakdown.

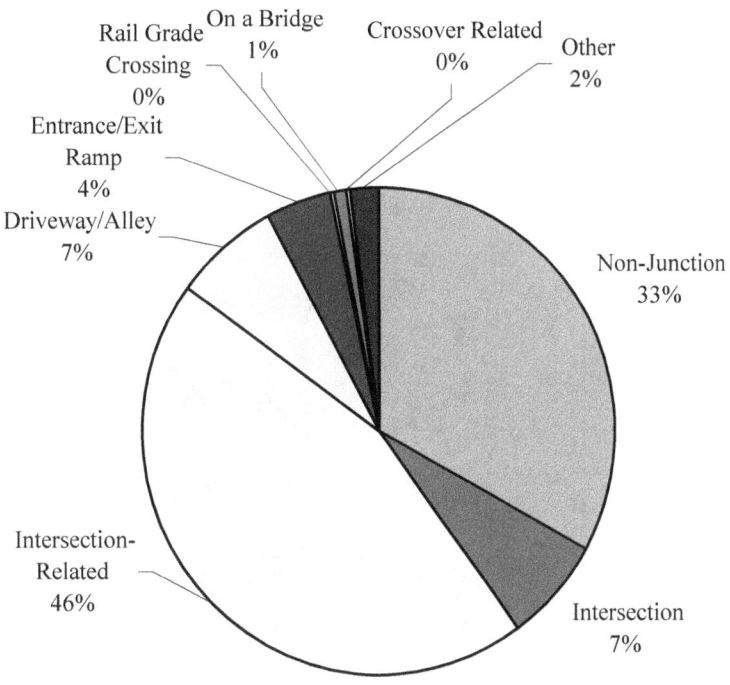

Figure 3-2(a). Distribution of Rear-end Crashes by Relation to Junction

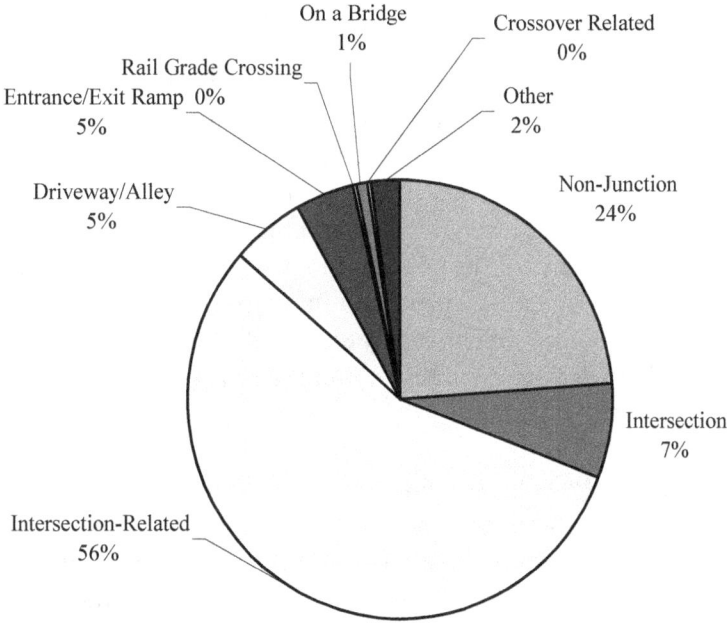

Figure 3-2(b). Distribution of Lead Vehicle Stopped Pre-Crash Scenario for Rear-end Crashes of Light Vehicles by Relation to Junction

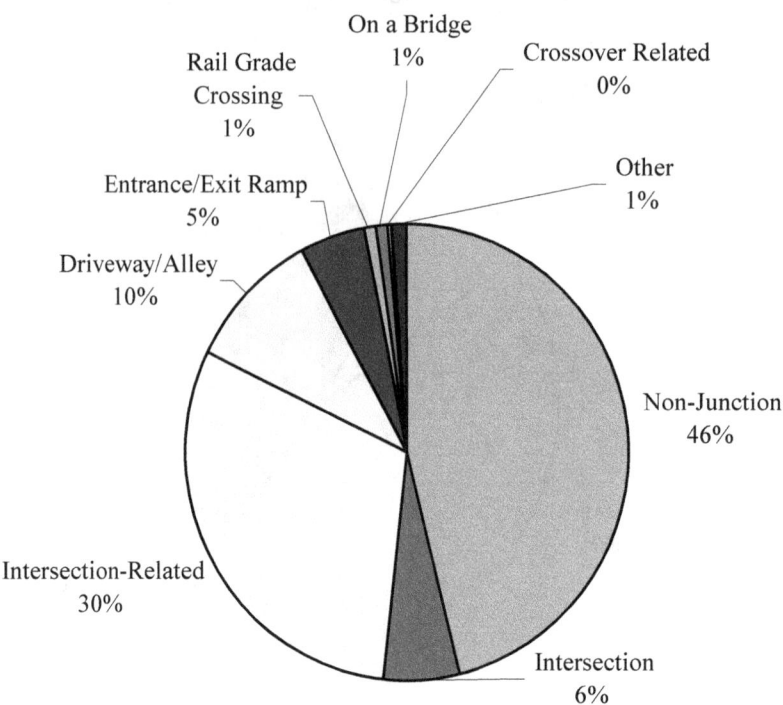

Figure 3-2(c). Distribution of Lead Vehicle Decelerating Pre-Crash Scenario for Rear-end Crashes of Light Vehicles by Relation to Junction

By way of comparison, the share of crashes for the next largest scenario (lead vehicle decelerating – Figure 3-2(c)) occurring at non-junction locations is almost twice the share for the previous scenario (46 percent), and the combined share occurring at junction locations is barely greater (46.2 percent, including 30.5 percent intersection-related, 5.7 percent intersection, and 10 percent driveway/alley). This suggests that a lead vehicle is about equally likely to be rear-ended while decelerating for reasons that have nothing to do with the proximity of an intersection (such as non-intersection-related congestion, or pulling into a roadside parking position, or slowing down to avoid striking another vehicle that is pulling out of a roadside parking position), as it is to be rear-ended while decelerating for intersection-related reasons such as slowing down to stop for a red light, stop sign, etc.

As observed in Section 2, there is a certain amount of overlap between the lead vehicle stopped and lead vehicle decelerating scenarios. In almost 52 percent of lead vehicle stopped crashes, the lead vehicle had come to a stop immediately prior to the crash. If these crashes were combined with the lead vehicle decelerating crashes into a common scenario, then the distribution with respect to relation to junction for this scenario is likely to be more weighted towards junction (i.e., intersection, intersection-related, and driveway/alley, combined) locations than the current lead vehicle decelerating scenario.

3.2. Crossing Path Crashes

Almost by definition, crossing path crashes are most likely to occur in the presence of crossing streams of traffic at a junction location (intersection, intersection-related, or driveway/alley). As seen clearly from Table 3-3 and Figure 3-3, 98 percent of light vehicle crossing path crashes occur at junction locations (73.8 percent at intersections, 4.4 percent at intersection-related locations, and 19.8 percent at junctions of driveways/alleys).

Table 3-3. Distribution of Pre-Crash Scenarios for Crossing Path Crashes of Light Vehicles by Relation to Junction (Two-vehicle crashes only; based on GES 2000)

RELATION TO JUNCTION	Left Turn Across Path/ Opposite Direction	Left Turn Into Path	Right Turn Into Path	Right Turn Across Path	Left Turn Across Path/ Lateral Direction	Straight Crossing Paths	Other/ Unknown	Total, All Scenarios
Non-Junction	3,000	1,000	1,000	-	1,000	3,000	2,000	10,000
Intersection	338,000	53,000	46,000	7,000	179,000	494,000	57,000	1,173,000
Intersection-Related	2,000	6,000	9,000	23,000	14,000	4,000	12,000	71,000
Driveway/Alley	77,000	31,000	36,000	4,000	108,000	40,000	19,000	314,000
Entrance/Exit Ramp	2,000	1,000	1,000	-	3,000	2,000	1,000	11,000
Rail Grade Crossing	-	-	-	-	-	-	-	-
On a Bridge	2,000	1,000	-	-	-	-	1,000	3,000
Crossover Related	1,000	2,000	-	-	-	1,000	-	4,000
Other	1,000	-	1,000	-	-	1,000	-	5,000
Total by scenario	425,000	94,000	93,000	34,000	306,000	545,000	93,000	1,590,000
Share by scenario	26.8%	5.9%	5.8%	2.1%	19.2%	34.3%	5.9%	100.0%
Non-Junction	0.6%	0.6%	0.6%	0.3%	0.3%	0.6%	1.9%	0.6%
Intersection	79.4%	56.0%	49.4%	19.1%	58.5%	90.6%	61.7%	73.8%
Intersection-Related	0.6%	6.5%	9.3%	68.3%	4.6%	0.7%	13.4%	4.4%
Driveway/Alley	18.1%	33.5%	38.5%	11.0%	35.3%	7.3%	20.1%	19.8%
Entrance/Exit Ramp	0.5%	0.8%	0.6%	0.5%	1.1%	0.5%	1.5%	0.7%
Rail Grade Crossing	0.0%	0.0%	0.0%	0.0%	0.0%	0.0%	0.0%	0.0%
On a Bridge	0.4%	0.6%	0.0%	0.0%	0.1%	0.0%	0.6%	0.2%
Crossover Related	0.1%	2.1%	0.3%	0.0%	0.0%	0.2%	0.3%	0.3%
Other	0.3%	0.0%	1.3%	0.8%	0.1%	0.2%	0.5%	0.3%
Combined Other Categories	1.3%	3.4%	2.2%	1.3%	1.4%	0.8%	2.9%	1.4%

Notes: Combined Other Categories is the sum of Entrance/Exit Ramp, Rail Grade Crossing, On a Bridge, Crossover Related, and Other categories.

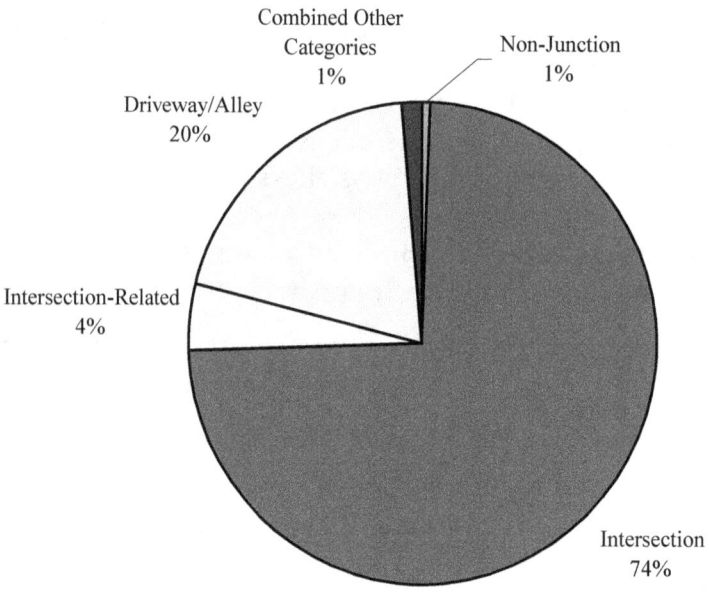

Figure 3-3. Distribution of Crossing Path Crashes of Light Vehicles by Relation to Junction

There is not much variation between the different pre-crash scenarios of crossing path crash scenarios in the distributions with respect to relation to junction. One interesting observation is the large shares of driveway/alley locations for the LTAP/LD, LTIP, and RTIP scenarios. An examination of Figure 2-4 shows why these scenarios are most likely to include a significant share of crashes involving vehicles pulling out of driveways.

3.3. Off-Roadway Crashes

Off-roadway crashes are classified into four subtypes, as described in Section 2.4. Table 3-4 and Figure 3-4 show the distribution of these subtypes with respect to relation to junction. The distribution for the entire crash type is, expectedly, very similar to the distribution for the largest subtype; i.e., single vehicle crashes, which constitute 88 percent of light vehicle off-roadway crashes. A majority of single vehicle off-roadway crashes (77.7 percent) occur at non-junction locations. This is, presumably, because many of the circumstances for loss of control (high speeds, especially on a curve) and road edge departure (fatigue/drowsiness, especially combined with poor visibility and roadway curvature), are more likely to occur at non-junction locations. The dominance of non-junction locations is even greater (95.5 percent) for no impact off-roadway crashes (fire, immersion, etc.)

Table 3-4. Distribution of Crash Subtypes for Off-Roadway Crashes of Light Vehicles by Relation to Junction (Based on GES 2000)

RELATION TO JUNCTION	Single Vehicle	Backing	No Impact	Multi-Vehicle	Total
Non-Junction	875,000	27,000	3,000	49,000	**954,000**
Intersection	7,000	-	-	2,000	**9,000**
Intersection-Related	150,000	3,000	-	12,000	**165,000**
Driveway/Alley	33,000	39,000	-	9,000	**81,000**
Entrance/Exit Ramp	39,000	-	-	3,000	**42,000**
Rail Grade Crossing	2,000	-	-	-	**2,000**
On a Bridge	17,000	-	-	3,000	**21,000**
Crossover Related	-	-	-	1,000	**2,000**
Other	2,000	1,000	-	1,000	**5,000**
Total by subtype	**1,126,000**	**70,000**	**3,000**	**81,000**	**1,280,000**
Share by subtype	**88.0%**	**5.5%**	**0.3%**	**6.3%**	**100.0%**
Non-Junction	77.7%	38.1%	95.5%	61.0%	**74.6%**
Intersection	0.6%	0.4%	0.0%	2.2%	**0.7%**
Intersection-Related	13.3%	4.0%	1.6%	14.8%	**12.9%**
Driveway/Alley	2.9%	55.4%	2.2%	11.6%	**6.3%**
Entrance/Exit Ramp	3.5%	0.0%	0.7%	3.7%	**3.3%**
Rail Grade Crossing	0.2%	0.0%	0.0%	0.0%	**0.1%**
On a Bridge	1.6%	0.5%	0.0%	3.6%	**1.6%**
Crossover Related	0.0%	0.0%	0.0%	1.4%	**0.1%**
Other	0.2%	1.6%	0.0%	1.6%	**0.4%**
Combined Other Categories	*2.0%*	*2.1%*	*0.0%*	*6.6%*	**2.3%**
Total	**100.0%**	**100.0%**	**100.0%**	**100.0%**	**100.0%**

Note: 1. This table, unlike the other tables in this section, shows the distribution of crash subtypes rather than pre-crash scenarios by relation to junction. A crash subtype is based on number of vehicles involved and critical event, rather than pre-crash sequence of events as described by the *Movement Prior to Critical Event* variable. Each crash subtype for off-roadway crashes can be further divided into pre-crash scenarios.
2. "Combined Other Categories" is the sum of Rail Grade Crossing, On a Bridge, Crossover Related, and Other categories.

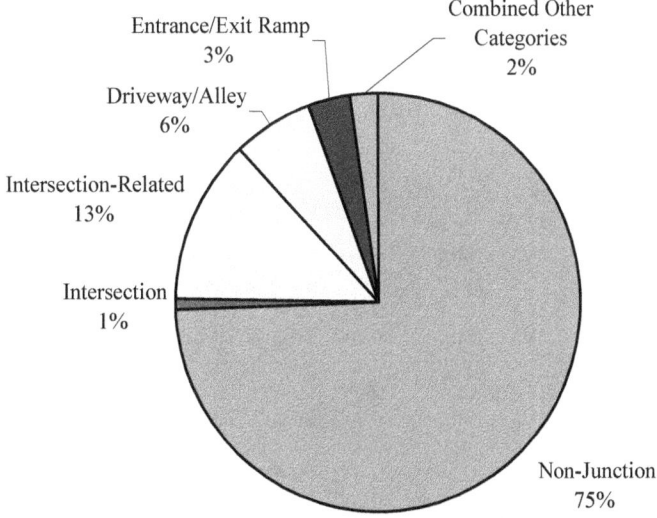

Figure 3-4. Distribution of Crash Subtypes by Relation to Junction for Off-Roadway Crashes of Light Vehicles

Multi-vehicle off-roadway crashes also occur more frequently at non-junction locations than anywhere else, but the dominance of non-junction locations is relatively less (61 percent). Junction locations comprise 28.6 percent of all multi-vehicle off-roadway crashes (2.2 percent intersections, 14.8 percent intersection-related, and 11.6 percent driveway/alley). By contrast, 55.4 percent of backing off-roadway crashes occur at driveway/alley locations; this is not surprising, considering that backing movements are common for driveways and alleys.

3.4. Lane Change Crashes

Table 2-6(a) shows a 9×7 matrix of possible pre-crash scenarios (i.e., 63 possible scenarios) for 2-vehicle lane change crashes. Of these scenarios, as many as 33 matrix cells are populated; i.e., 33 of the 63 scenarios are actually found to exist in any significant numbers in the GES data. For simplicity, this study did not attempt to find the distribution with respect to relation to junction for all these 33 scenarios. Instead, the distribution was found for the seven most common scenarios, as well as the residual "All Other Combinations" scenario, in Table 2-6(b).

Overall, 44.9 percent of all light vehicle lane change crashes occur at non-junction locations, as listed in Table 3-5 and Figure 3-5(a). On the other hand, about 49.4 percent of these crashes happen at junction locations (18.8 percent intersection, 19.1 percent intersection-related, and 11.5 percent driveway/alley). This overall distribution is not very significant because it is an average of several dissimilar distributions for the different scenarios.

Table 3-5. Distribution of Pre-Crash Scenarios for Lane Change Crashes of Light Vehicles by Relation to Junction (Two-vehicle crashes only; based on GES 2000)

RELATION TO JUNCTION	Scenario Number							All Other	Total, All Scenarios
	1	2	3	4	5	6	7		
Non-Junction	115,000	2,000	23,000	-	11,000	11,000	14,000	69,000	**245,000**
Intersection	14,000	41,000	6,000	11,000	3,000	1,000	2,000	25,000	**103,000**
Intersection-Related	45,000	10,000	11,000	3,000	5,000	5,000	3,000	22,000	**104,000**
Driveway/Alley	6,000	21,000	1,000	17,000	1,000	2,000	1,000	14,000	**63,000**
Entrance/Exit Ramp	6,000	1,000	1,000	-	1,000	1,000	-	9,000	**19,000**
Rail Grade Crossing	-	-	-	-	-	-	-	-	-
On a Bridge	2,000	-	-	1,000	1,000	-	-	2,000	**6,000**
Crossover Related	1,000	-	-	-	-	-	-	-	**1,000**
Other	2,000	-	-	-	-	-	-	1,000	**5,000**
Total by scenario	**190,000**	**76,000**	**42,000**	**32,000**	**22,000**	**21,000**	**20,000**	**142,000**	**545,000**
Non-Junction	60.7%	3.1%	54.8%	1.4%	50.6%	52.8%	71.3%	48.6%	**44.9%**
Intersection	7.2%	53.6%	13.3%	34.6%	14.8%	6.0%	10.8%	17.6%	**18.8%**
Intersection-Related	23.6%	13.4%	26.0%	9.8%	22.4%	25.4%	14.0%	15.5%	**19.1%**
Driveway/Alley	3.1%	27.2%	3.3%	51.3%	4.2%	8.9%	2.7%	9.9%	**11.5%**
Entrance/Exit Ramp	2.9%	1.1%	1.5%	0.0%	4.2%	2.7%	0.0%	6.3%	**3.4%**
Rail Grade Crossing	0.0%	0.0%	0.0%	0.0%	0.0%	1.2%	0.0%	0.0%	**0.0%**
On a Bridge	1.0%	0.7%	0.1%	1.6%	2.8%	1.2%	0.0%	1.4%	**1.0%**
Crossover Related	0.3%	0.4%	0.6%	0.0%	0.0%	0.2%	0.0%	0.0%	**0.2%**
Other	1.2%	0.5%	0.2%	1.2%	1.1%	1.5%	1.3%	0.7%	**0.9%**
Combined Other Categories	*2.5%*	*1.6%*	*0.9%*	*2.9%*	*3.9%*	*4.1%*	*1.3%*	*2.1%*	*2.3%*

Note: "Combined Other Categories" is the sum of Rail Grade Crossing, On a Bridge, Crossover Related, and Other categories. For an explanation of pre-crash scenarios, see Tables 2.6(a) and 2.6(b), and Figure 2-6.

The most frequent scenario that involves 1 vehicle going straight and another executing a simple lane change maneuver occurs at non-junction locations in 60.7 percent of the cases, as shown in Figure 3-5(b). This is expected because most drivers are likely to change lanes in between intersections. However, a fairly large share (33.9 percent) of simple lane change crashes do occur at or close to junctions: 7.2 percent intersection, 23.6 percent intersection-related, and 3.1 percent driveway/alley. A likely explanation for these crashes is that the driver of the lane-changing vehicle made a last-second lane change to get into the turning lane in order to make a turn at the intersection, or to the right lane in order to pull into a driveway or alley.

Figure 3-5(c) illustrates the distribution of the second most frequent scenario by relation to junction, which consists of one vehicle going straight and another turning. The majority of these crashes (94.2 percent) is reported at junction locations: 53.6 percent intersection, 13.4 percent intersection-related, and 27.2 percent driveway/alley. This result is expected given the definition of this scenario. For the remaining 6 percent of these crashes, it is possible that either the relation to junction or the vehicle movement was miscoded; i.e., it is possible that a lane change/merge maneuver was miscoded as turning in the GES for the 1.1 percent of these crashes reported at entrance/exit ramps.

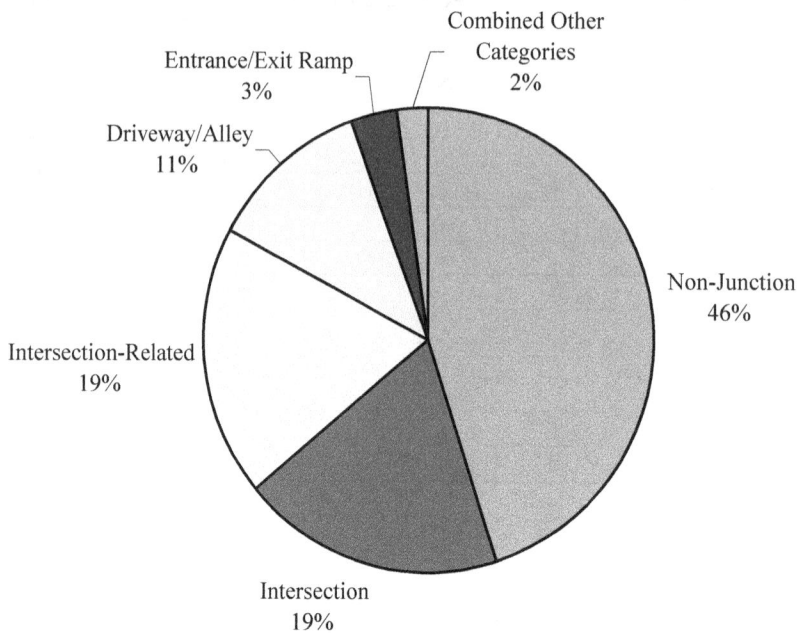

Figure 3.5(a). Distribution of Lane Change Crashes of Light Vehicles by Relation to Junction

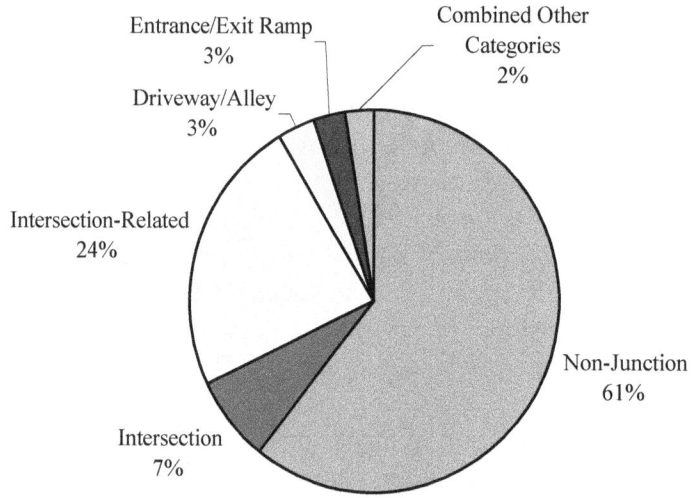

Figure 3.5(b). Distribution of Pre-crash Scenario 1 (One Vehicle Going Straight, One Vehicle Executing Simple Lane Change) for Lane Change Crashes of Light Vehicles by Relation to Junction

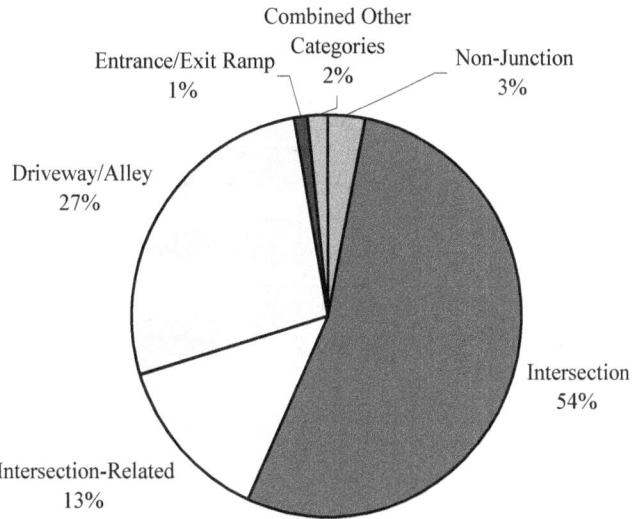

Figure 3-5(c). Distribution of Pre-Crash Scenario 2 (One Vehicle Going Straight, One Vehicle Turning) for Lane Change Crashes of Light Vehicles by Relation to Junction

3.5. Animal Crashes

The overwhelming majority of light vehicle crashes with animals (almost 96 percent) occur at non-junction locations, as shown in Table 3-6 and Figure 3-6. This is unlikely to reflect anything significant other than the fact that struck animals (mostly deer) roam outside populated areas and thus, are away from junctions.

Table 3-6. Distribution of Pre-Crash Scenarios for Crashes of Light Vehicles with Animals by Relation to Junction (single-vehicle crashes only; based on GES 2000)

RELATION TO JUNCTION	Vehicle Going Straight - Animal in Roadway	Vehicle Negotiating a Curve - Animal in Roadway	Other	Total, All Scenarios
Non-Junction	225,000	8,000	4,000	237,000
Intersection	2,000	-	-	2,000
Intersection-Related	4,000	-	1,000	5,000
Driveway/Alley	-	-	-	-
Entrance/Exit Ramp	1,000	1,000	-	2,000
Rail Grade Crossing	-	-	-	-
On a Bridge	1,000	-	-	1,000
Crossover Related	-	-	-	-
Other				
Total by scenario	232,000	9,000	6,000	247,000
Share by scenario	93.8%	3.8%	2.3%	100.0%
Non-Junction	96.9%	86.3%	69.9%	95.9%
Intersection	0.7%	0.0%	1.3%	0.7%
Intersection-Related	1.6%	0.0%	19.3%	1.9%
Driveway/Alley	0.0%	0.0%	2.2%	0.1%
Entrance/Exit Ramp	0.3%	12.8%	1.4%	0.9%
Rail Grade Crossing	0.0%	0.0%	0.0%	0.0%
On a Bridge	0.4%	0.8%	1.3%	0.4%
Crossover Related	0.0%	0.0%	0.0%	0.0%
Other	0.1%	0.0%	4.6%	0.2%
Combined Other Categories	0.5%	0.8%	5.9%	0.6%

Note: "Combined Other Categories" is the sum of Rail Grade Crossing, On a Bridge, Crossover Related, and Other categories.

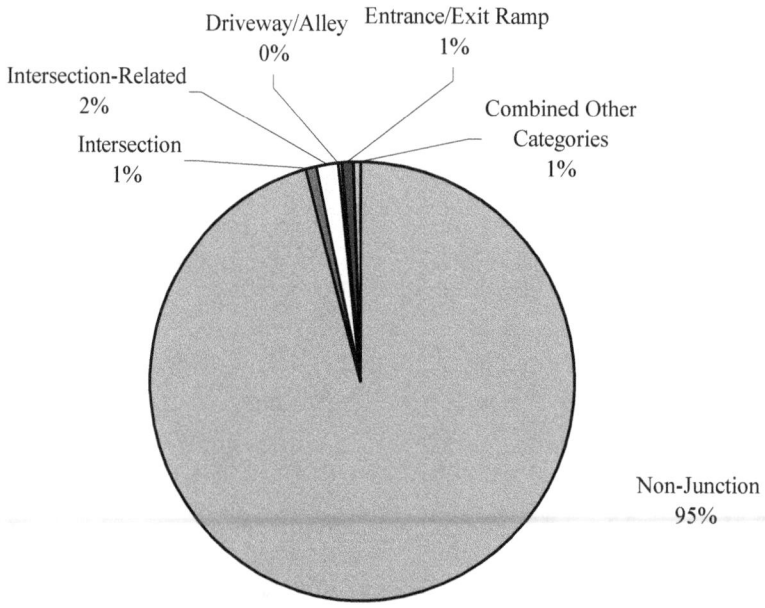

Figure 3-6. Distribution of Crashes of Light Vehicles with Animals by Relation to Junction

3.6. Opposite Direction Crashes

Table 3-7 and Figure 3-7(a) show that a large majority of light vehicle opposite direction crashes, 80.6 percent, occurs at non-junction locations. The share of non-junction locations for individual pre-crash scenarios varies from a high of 93 percent for the negotiating a curve/encroaching scenario to a low of 70.4 percent for the vehicle failure scenario.

The two largest scenarios, going straight/encroaching and negotiating a curve/encroaching, are selected for further analysis as illustrated respectively in Figure 3-7(b) and 3.7(c). About 71.6 percent of the going straight/encroaching scenario crashes occurred at non-junctions, which is below the average of 80.6 percent for the opposite direction crash type. In contrast, 93 percent of all negotiating a curve/encroaching scenario crashes happened at non-junctions, which corresponds to the largest non-junction share among all opposite direction pre-crash scenarios.

Table 3-7. Distribution of Pre-Crash Scenarios for Opposite Direction Crashes of Light Vehicles by Relation to Junction (two-vehicle crashes only; based on GES 2000)

RELATION TO JUNCTION	Going Straight/ Encroaching	Going Straight/ In Lane	Negotiating a Curve/Encroaching	Negotiating a Curve/ In Lane	Control Loss (Except when Passing)	Passing	Vehicle Failure	Other	Total, All Scenarios
Non-Junction	53,000	5,000	55,000	3,000	1,000	4,000	1,000	10,000	132,000
Intersection	9,000	1,000	1,000	-	-	-	-	-	11,000
Intersection-Related	9,000	1,000	2,000	-	-	1,000	-	-	14,000
Driveway/Alley	1,000	-	1,000	-	-	-	-	1,000	3,000
Entrance/Exit Ramp	-	-	-	-	-	-	-	-	-
Rail Grade Crossing	-	-	-	-	-	-	-	-	-
On a Bridge	1,000	-	-	-	-	-	-	-	2,000
Crossover Related	-	-	-	-	-	-	-	-	-
Other	-	-	-	-	-	-	-	-	1,000
Total by scenario	74,000	7,000	59,000	3,000	1,000	5,000	1,000	12,000	163,000
Share by scenario	**45.4%**	**4.6%**	**36.4%**	**1.8%**	**0.7%**	**3.1%**	**0.9%**	**7.2%**	**100.0%**
Non-Junction	71.6%	70.7%	93.0%	89.6%	80.5%	75.8%	70.4%	81.7%	**80.6%**
Intersection	11.7%	6.9%	2.1%	0.0%	7.0%	4.8%	0.0%	3.8%	**6.9%**
Intersection-Related	12.7%	16.4%	2.7%	9.6%	3.8%	17.7%	6.0%	2.8%	**8.5%**
Driveway/Alley	1.4%	4.4%	1.3%	0.0%	3.4%	0.1%	19.2%	5.7%	**1.9%**
Entrance/Exit Ramp	0.5%	0.0%	0.1%	0.8%	0.0%	0.0%	0.0%	0.1%	**0.3%**
Rail Grade Crossing	0.1%	0.0%	0.0%	0.0%	0.0%	0.0%	0.0%	0.0%	**0.0%**
On a Bridge	1.3%	1.7%	0.3%	0.0%	5.3%	1.6%	4.3%	3.1%	**1.1%**
Crossover Related	0.5%	0.0%	0.1%	0.0%	0.0%	0.0%	0.0%	0.0%	**0.3%**
Other	0.4%	0.0%	0.3%	0.0%	0.0%	0.0%	0.0%	2.8%	**0.5%**
Combined Other Categories	*2.2%*	*1.7%*	*0.7%*	*0.0%*	*5.3%*	*1.6%*	*4.3%*	*5.9%*	*1.9%*

Note: "Combined Other Categories" is the sum of Rail Grade Crossing, On a Bridge, Crossover Related, and Other categories.

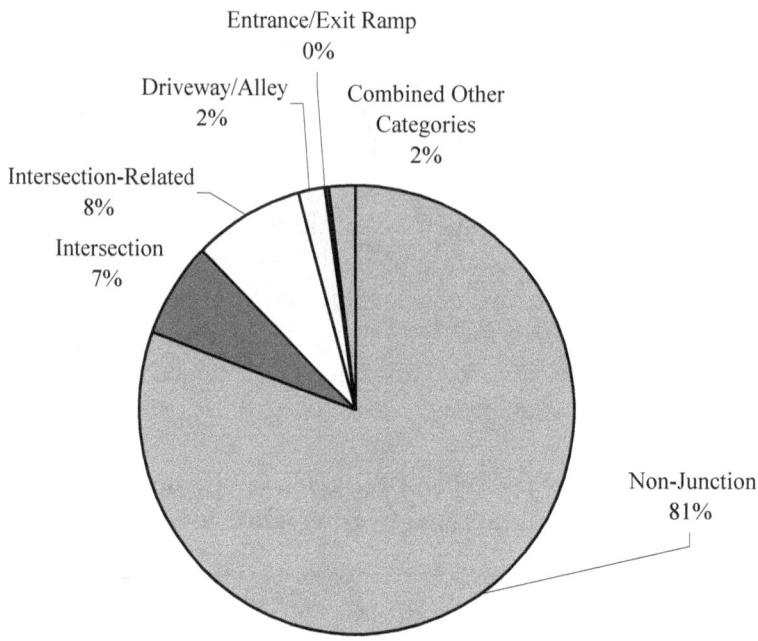

Figure 3-7(a). Distribution of Opposite Direction Crashes of Light Vehicles by Relation to Junction

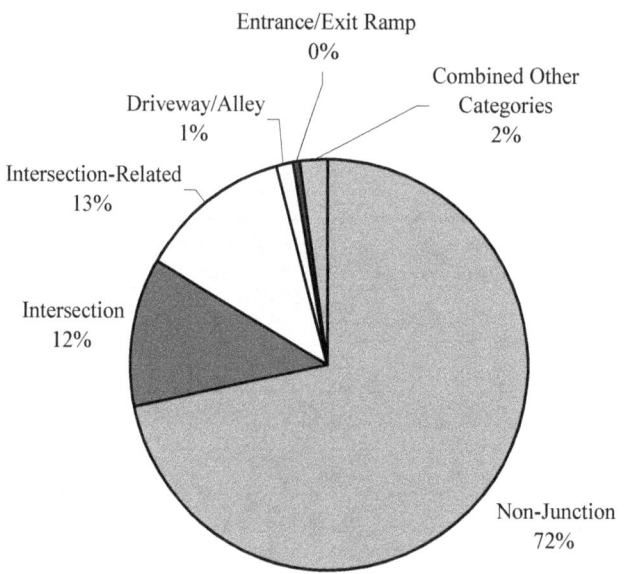

Figure 3-7(b). Distribution of Going Straight/Encroaching Pre-Crash Scenario for Opposite Direction Crashes of Light Vehicles by Relation to Junction

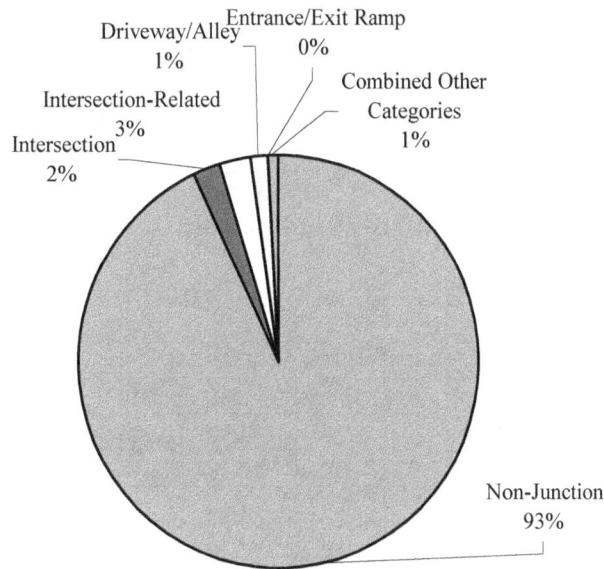

Figure 3-7(c). Distribution of Negotiating a Curve/Encroaching Pre-Crash Scenario for Opposite Direction Crashes of Light Vehicles by Relation to Junction

The higher non-junction share of the negotiating a curve/encroaching scenario may indicate that, in roadway design, it is rare to have intersections and driveways on road segments with significant curvature. Therefore, it follows that opposite direction crashes in which one vehicle encroached into the lane in the opposite direction while following a curve are more likely to occur on road segments that are not at, or close to, intersections. On the other hand, the junction share of going straight/encroaching crashes is significant (25.8 percent), consisting of: intersection (11.7 percent), intersection-related (12.7 percent), and driveway/alley (1.4 percent). This may be attributed to encroachment into the opposite direction at an intersection location, but it may also be attributed to possible miscoding. It is conceivable that some right turn across path/lateral direction crashes, as well as some left turn across path/opposite direction crashes (see Figure 2-4), are miscoded as opposite direction crashes, with the vehicle movement of the turning vehicle miscoded as "going straight."

3.7. Pedestrian Crashes

Table 3-8 and Figure 3-8 demonstrate that the majority of light vehicle crashes with pedestrians occurs near junctions, as evident by the 62.4 percent of these crashes reported at intersections (31.3 percent), intersection-related roadways (26.2 percent), driveways/alleyways (3.8 percent), and entrance/exit ramps (1.1 percent). Most of the remaining crashes occur away from junctions (37.0 percent).

The most dominant scenario of pedestrian crashes – vehicle going straight and pedestrian crossing the road – occurs almost equally at intersections (31.7 percent), intersection-related roadways (33.3 percent), and away from junctions (33.3 percent). On the other hand, almost two-thirds of crashes preceded by the second most frequent scenario – vehicle going straight and pedestrian darting onto the road – happen away from junctions. As expected, most pedestrian crashes with light vehicles turning are reported to occur at intersections and intersection-related roadways (95-98 percent). About 70 percent of pedestrians struck while walking along the road were away from junctions. Similarly, the majority of light vehicle crashes with pedestrians playing or working in the road (63 percent) happens away from junctions.

Table 3-8. Distribution of Pre-Crash Scenarios for Pedestrian Crashes of Light Vehicles by Relation to Junction (Based on GES 2000)

Relation to Junction	Scenarios									Total
	1	2	3	4	5	6	7	8	9	
Non-Junction	33.3%	66.0%	2.4%	0.5%	48.4%	70.1%	63.1%	44.6%	28.3%	37.0%
Intersection	31.7%	13.8%	58.9%	67.7%	23.8%	7.7%	16.5%	1.6%	40.3%	31.3%
Intersection-Related	33.3%	16.6%	35.8%	30.2%	16.2%	7.0%	16.2%	22.5%	15.1%	26.2%
Driveway/Alley	1.5%	2.7%	3.0%	1.6%	0.6%	1.4%	4.3%	31.3%	15.3%	3.8%
Entrance/Exit Ramp	0.0%	0.1%	0.0%	0.0%	5.2%	13.8%	0.0%	0.0%	0.0%	1.1%
Rail Grade Crossing	0.1%	0.2%	0.0%	0.0%	0.0%	0.0%	0.0%	0.0%	0.0%	0.1%
On a Bridge	0.0%	0.5%	0.0%	0.0%	5.1%	0.0%	0.0%	0.0%	0.0%	0.4%
Crossover Related	0.0%	0.0%	0.0%	0.0%	0.0%	0.0%	0.0%	0.0%	0.0%	0.0%
Other	0.0%	0.0%	0.0%	0.0%	0.7%	0.0%	0.0%	0.0%	1.0%	0.1%
Total	100.0%	100.0%	100.0%	100.0%	100.0%	100.0%	100.0%	100.0%	100.0%	100.0%

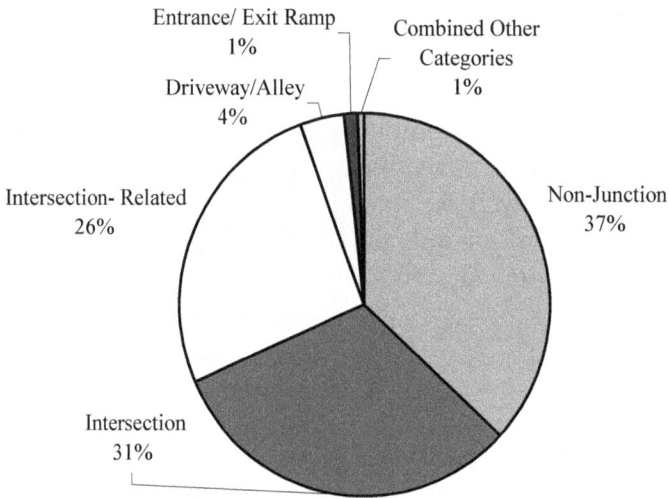

Figure 3-8. Distribution of Pedestrian Crashes of Light Vehicles by Relation to Junction

3.8. Pedalcyclist Crashes

Table 3-9 and Figure 3-9(a) indicate that 82.7 percent of all light vehicle crashes with pedalcyclists occur in and around junctions: 45.2 percent at intersections, 22.8 percent intersection-related, and 14.7 percent near driveways/alleys. This is an expected result because, as mentioned in Section 2.10, many pre-crash scenarios are similar to the scenarios for crossing path crashes illustrated in Figure 2-4. It is not surprising that interactions between vehicles and pedalcyclists leading to crashes are more likely to occur at intersections and related locations, with crossing streams of traffic.

Table 3-9. Distribution of Pre-Crash Scenarios for Pedalcyclist Crashes of Light Vehicles by Relation to Junction (Based on GES 2000)

Relation to Junction	Vehicle Traveling Straight/Crossing Paths	Vehicle Traveling Straight/Parallel Paths	Vehicle Turning Right/Crossing Paths	Vehicle Turning Right/Parallel Paths	Vehicle Turning Left/Crossing Paths	Vehicle Turning Left/Parallel Paths	Vehicle Starting in Traffic Lane/Crossing Paths	Other/Unknown	Total
Non-Junction	2,000	4,000	-	-	-	-	-	1,000	7,000
Intersection	12,000	1,000	3,000	1,000	1,000	2,000	1,000	1,000	22,000
Intersection-Related	3,000	1,000	2,000	1,000	1,000	1,000	1,000	1,000	11,000
Driveway/Alley	3,000	-	-	1,000	1,000	1,000	-	1,000	7,000
Entrance/Exit Ramp	-	-	-	-	-	-	-	-	-
Rail Grade Crossing	-	-	-	-	-	-	-	-	-
On a Bridge	-	-	-	-	-	-	-	-	-
Crossover Related	-	-	-	-	-	-	-	-	-
Other	-	-	-	-	-	-	-	-	-
Total by scenario	**21,000**	**7,000**	**5,000**	**3,000**	**3,000**	**3,000**	**2,000**	**3,000**	**47,000**
Share by scenario	**44.7%**	**14.9%**	**10.6%**	**6.4%**	**6.4%**	**6.4%**	**4.3%**	**6.4%**	**100.0%**
Non-Junction	10.7%	62.5%	0.0%	0.0%	0.0%	0.0%	0.0%	23.2%	15.5%
Intersection	57.5%	10.1%	49.3%	42.0%	43.4%	51.0%	64.8%	21.8%	45.2%
Intersection-Related	16.7%	20.4%	40.6%	32.6%	23.3%	25.7%	22.8%	26.5%	22.8%
Driveway/Alley	13.5%	5.7%	7.8%	23.0%	31.5%	22.6%	11.0%	25.2%	14.7%
Entrance/Exit Ramp	0.0%	0.7%	0.9%	2.4%	1.8%	0.7%	0.0%	1.2%	0.6%
Rail Grade Crossing	0.0%	0.0%	0.0%	0.0%	0.0%	0.0%	0.0%	0.7%	0.1%
On a Bridge	1.5%	0.2%	0.0%	0.0%	0.0%	0.0%	0.0%	0.7%	0.7%
Crossover Related	0.0%	0.0%	0.0%	0.0%	0.0%	0.0%	0.0%	0.0%	0.0%
Other	0.3%	0.3%	1.4%	0.0%	0.0%	0.0%	1.4%	0.7%	0.4%
Combined Other Categories	*1.8%*	*1.5%*	*1.4%*	*0.0%*	*0.0%*	*0.0%*	*1.4%*	*2.1%*	**1.2%**

Note: "Combined Other Categories" is the sum of Rail Grade Crossing, On a Bridge, Crossover Related, and Other categories.

Pedalcyclist crashes involving the vehicle turning left across the path of the cyclist from a parallel direction are reported almost entirely at junction locations, as seen in Table 3-9 and Figure 3-9(b). In contrast, 62.5 percent of all crashes for the cyclist traveling straight on a parallel path with the vehicle occur at non-junction locations. These are probably cases in which a pedalcyclist or vehicle was attempting to change lanes. The vehicle turning right into the path of cyclist from parallel direction scenario, however, occurs almost 100 percent of the time at junction locations (42 percent at intersections, 32.6 percent intersection-related, 23 percent driveways/alleys). These are cases of vehicles making a right turn at an intersection, or trying to pull into a driveway or alley, encroaching into the pedalcyclists' lane (which may be a dedicated bicycle lane along the right side of the road, as mentioned earlier).

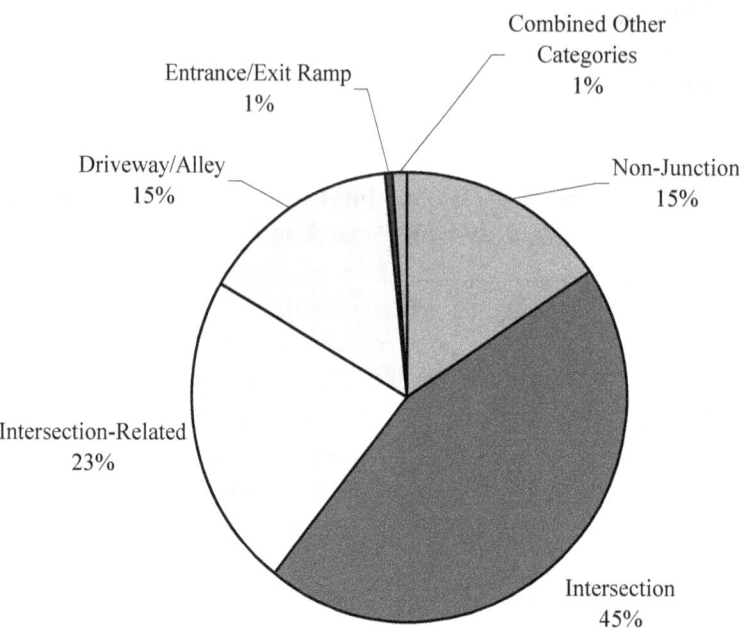

Figure 3-9(a). Distribution of Pedalcyclist Crashes of Light Vehicles by Relation to Junction

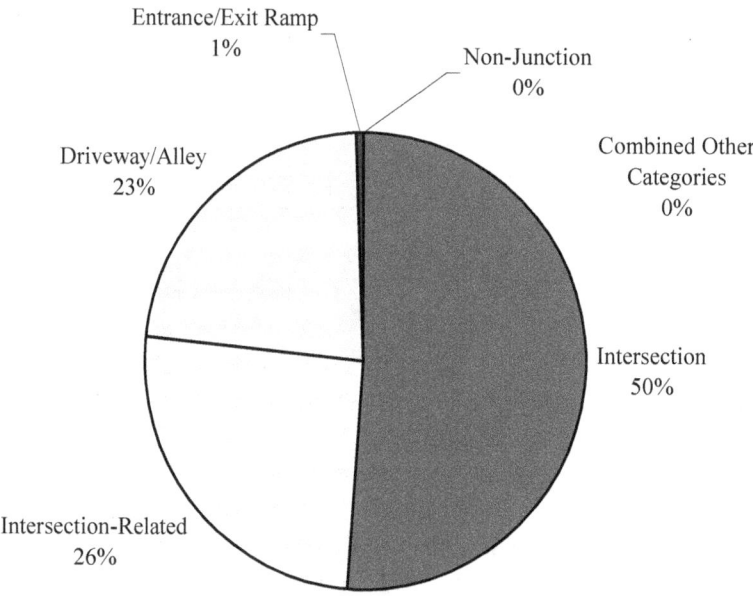

Figure 3-9(b). Distribution of Vehicle Turning Left/Parallel Paths Pre-Crash Scenario for Pedalcyclist Crashes of Light Vehicles by Relation to Junction

3.9. Light Vehicle Crash Distribution by Traffic Control Device

Table 3-10 presents a distribution of relation to junction by traffic control device for all PR light vehicle crashes based on 2000 GES data. The *Traffic Control Device* variable in the GES identifies the presence and type of the traffic control device. This variable applies at the crash level, not at the vehicle level (at the time of this analysis, the GES was being revised to incorporate the traffic control device at the vehicle level in the year 2000 and beyond). That is, if several types of controls were present at a junction, then the control device with the lowest number was coded for the entire crash. According to the GES coding manual, the *Traffic Control Signal (On Colors)* code is used if the police accident report indicates a signal that processes through the green, amber, and red times. A *Stop Sign* is coded in the GES if there is at least one stop sign present at an intersection or driveway. The stop sign takes precedence over other signs such as a "yield" sign. It should be noted that the GES does not provide information on whether an intersection or a driveway is 2-way or 4-way stop sign controlled. *No Controls* is coded in the GES if at the time of the crash there was no intent to control (regulate or warn) vehicle traffic (i.e., an uncontrolled intersection). This code is also used if statutory controls apply (e.g., state law requires that when two vehicles meet at an uncontrolled intersection, the one on the right has the right-of-way). As seen in Table 3-10, about 62 percent of all PR light vehicle crashes occur on roadways with no controls. Signal- and stop sign-controlled junctions account for about 32 percent of all PR light vehicle crashes.

Table 3-10. Distribution of PR Light Vehicle Crashes by Relation to Junction by Traffic Control Device (Based on 2000 GES)

Relation to Junction	Signal	Stop Sign	No Controls	Other	Unknown	Total	
Non-Junction	-	7,000	2,476,000	100,000	13,000	2,595,000	40.6%
Intersection	669,000	483,000	286,000	79,000	-	1,518,000	23.8%
Intersection-Related	626,000	161,000	437,000	65,000	1,000	1,289,000	20.2%
Driveway/Alley	31,000	31,000	596,000	18,000	-	677,000	10.6%
Entrance/Exit Ramp	20,000	13,000	101,000	29,000	1,000	164,000	2.6%
Rail Grade Crossing	-	-	3,000	13,000	-	15,000	0.2%
On a Bridge	4,000	-	48,000	4,000	1,000	57,000	0.9%
Crossover Related	2,000	2,000	13,000	2,000	-	19,000	0.3%
Other	4,000	1,000	26,000	23,000	-	55,000	0.9%
Total	1,357,000	699,000	3,984,000	333,000	16,000	6,389,000	100.0%
	21.2%	10.9%	62.4%	5.2%	0.3%	100.0%	

About 1,277,000 (46.4 percent) and 637,000 (23.1 percent) of all PR light vehicle crashes at intersections and intersection-related roadways happened, respectively, under signal and stop sign controls, as indicated in Table 3-11. A total of 695,000 (25.3 percent) of these crashes were reported at uncontrolled intersections. Signalized intersections and intersection-related roadways experienced 524,000 rear-end crashes (57.7 percent), 104,000 lane change crashes (49.1 percent), 18,000 pedestrian crashes (46.1 percent), 544,000 crossing path crashes (43.7 percent), and 19,000 backing crashes (39.1 percent), based on 2000 GES statistics. About 12,000 or 38.0

percent of all pedalcyclist crashes at intersections and intersection-related roadways occurred in the presence of stop signs. Most animal, off-roadway, and opposite direction crashes at intersections and intersection-related roadways were reported under no controls – 5,000 animal (84.1 percent), 100,000 off-roadway (57.7 percent), and 14,000 opposite direction (55.5 percent) crashes.

Table 3-11. Distribution of Crash Type by Relation to Junction by Traffic Control Device

Crash Type	Relative Frequency at Intersection & Intersection-Related	Traffic Control Device				Total
		Signal	Stop Sign	No Controls	Other/Unk.	
Off-Roadway	13.6%	15.5%	22.5%	57.7%	4.3%	100.0%
Pedestrian	57.5%	46.1%	16.7%	34.0%	3.1%	100.0%
Pedalcyclist	68.0%	32.9%	38.0%	24.8%	4.3%	100.0%
Rear-End	50.3%	57.7%	7.5%	29.0%	5.8%	100.0%
Lane Change	37.7%	49.1%	4.6%	41.8%	4.6%	100.0%
Crossing Paths	78.2%	43.7%	37.9%	13.3%	5.2%	100.0%
Opposite Direction	15.3%	35.2%	5.7%	55.5%	3.6%	100.0%
Backing	37.6%	39.1%	34.4%	24.1%	2.4%	100.0%
Animal	2.6%	6.3%	4.6%	84.1%	5.0%	100.0%
Other	26.4%	34.8%	19.0%	39.6%	6.6%	100.0%
Total		46.4%	23.1%	25.3%	5.2%	100.0%

4. CRASH TAXONOMY BASED ON PRE-CRASH SCENARIOS

This section builds a new crash taxonomy based on pre-crash scenarios by transforming the typology of major crash types into a set of pre-crash scenarios that cut across different crash types. By definition, pre-crash scenarios combine vehicle movements and maneuvers with critical events that occur immediately prior to a collision. It is noteworthy that certain combinations of vehicle movements/maneuvers and critical events lead to more than one crash type. For example, a catastrophic loss of vehicle control due to excessive speed could lead to a vehicle running off the roadway, rear-ending another vehicle in front of it, or encroaching into another lane and leading to a lane-change type crash. From a crash avoidance perspective, the problem of vehicle control loss is identical in all three cases. A potential crash countermeasure function would detect the excessive speed or the imminent loss of control regardless of what crash type these conditions might lead to. Therefore, pre-crash scenario taxonomy would be of great practical benefit since crash countermeasure systems are designed to detect crash-imminent situations and take preventive action.

4.1. Development of Pre-Crash Scenario Taxonomy

The results of Section 2 help create a detailed matrix that correlates pre-crash scenarios by major crash types. Table 4-1 presents a condensed version of this matrix, showing only the pre-crash scenarios with an individual frequency of at least 100,000 crashes based on 2000 GES estimates. The adoption of this crash frequency criterion disqualifies the inclusion of any backing and pedestrian pre-crash scenarios. Pedestrian crashes are unique among crash types since none of the concomitant pre-crash scenarios resemble other scenarios in any of the other major crash types. Pedalcyclist crashes are counted in Table 4-1 because they involve non-motorized vehicles that use the same roadway as motor vehicles (i.e., move along with traffic – unlike pedestrians), and often exhibit the same combination of movements as two-vehicle crashes, especially crossing path crashes. It should be noted that the inclusion of pedalcyclist crashes in Table 4-1 does not change any of the scenario rankings.

Table 4-1. Distribution of Top Pre-Crash Scenarios for Light Vehicle Crashes by Crash Type (Based on GES 2000)

Pre-Crash Scenario	Crash Type[2]							
	Rear-End	Crossing Paths	Single Vehicle Run-Off-Road[3]	Lane Change[4]	Animal	Opposite Direction	Pedalcyclist[5]	Total, All Crash Types
Lead Vehicle Stopped	888,000	-	[6]	-	-	-	-	888,000
Straight Crossing Paths[1]	-	536,000	-	-	-	-	21,000	557,000
Control Loss	8,000	12,000	436,000	29,000	-	1,000	-	486,000
Left Turn Across Path/ Opposite Direction	-	424,000	-	-	-	-	[8]	424,000
Drifting (Going Straight)	7,000	-	281,000	38,000	-	74,000	-	400,000
Lead Vehicle Decelerating	397,000	-	[6]	-	-	-	-	397,000
Left Turn Across Path/ Lateral Direction	-	304,000	-	-	-	-	[9]	304,000
Vehicle Changing Lanes	55,000	-	29,000	190,000	-	-	-	274,000
Vehicle going straight/ animal in roadway	-	-	[7]	-	232,000	-	-	232,000
Drifting (Negotiating a Curve)	-	-	110,000	4,000	-	60,000	-	174,000
Lead vehicle moving at constant speed	139,000	-	[6]	-	-	-	-	139,000
All other scenarios	18,000	220,000	270,000	284,000	15,000	28,000	21,000	856,000

Notes:
[1] Note that the 557,000 Straight Crossing Paths crashes include 21,000 pedalcyclist crashes. If pedalcyclist crashes are excluded, this scenario remains the second largest.
[2] Only selected crash types are shown. Pedestrian crashes have a unique set of scenarios. Backing crashes do not include any scenario with a total over 100,00. Hence, these two crash types are not shown in the table. Rear-end, lane change, and opposite direction crashes involving more than two vehicles, as well as run-off-road, animal, and pedalcyclist crashes involving more than one motor vehicle, are not included in the counts. Therefore, the total of the All Crash Types column is less than the total number of police-reported crashes involving light vehicles in 2000.
[3] The "control loss" scenario for this crash type is a sum of four scenarios (going straight and lost control, negotiating a curve and lost control, initiating a maneuver and lost control, and other control loss) from Table 2-5(b). See explanation in Chapter 4 as to why these scenarios were combined for the purpose of the analysis.
[4] The lane change scenarios are somewhat different from Table 2-6(a) and Table 2-6(b). Specifically, the control loss scenario is the sum of all types of control loss, regardless of the combination of vehicle movements involved. In addition, the drifting scenario is further differentiated by whether drifting occurred on a straight road or a curve.
[5] There are arguments for treating pedalcyclist crashes as a unique category along with pedestrian crashes. However, unlike pedestrian crashes, pedalcyclist crashes involve non-motorized vehicles which use the same roadway, i.e. move along with vehicular traffic, and often involve the same combination of movements as two-vehicle crashes, especially crossing paths and lane change crashes.
[6] A total of 18,000 run-off-the-road crashes resulted from maneuvers to avoid potential rear-end crashes. The particular scenarios for the rear-end crashes (lead vehicle stopped, lead vehicle decelerating, or lead vehicle moving at constant speed, etc.) are not known.
[7] A total of 39,000 run-off-the-road crashes resulted from maneuvers to avoid animal/object collisions. Note that it is not known how many of these were to avoid animal collisions and how many were to avoid object collisions; also, it is not known if these occurred on a straight road or a curve. Therefore, a precise number cannot be entered for this cell.
[8] A total of 2,000 pedalcyclist crashes occurred in which the vehicle was turning left on a parallel path with the pedalcyclist. Note that it is not known in how many of these crashes the pedalcyclist was traveling is the opposite direction.
[9] A total of 3,000 known pedalcyclist crashes occurred in which the vehicle was turning left on a crossing path with the pedalcyclist. Note that it is not known in how many of these crashes the vehicle turned left across the path of the pedalcyclist.

The sum of the "Total, All Crash Types" column in Table 4-1 is 5,131,000 PR light vehicle crashes, which is less than the combined sum of 5,699,000 crashes of the 7 crash types shown. This lower amount is due to the exclusion of

- Rear-end, lane change, and opposite direction crashes involving more than two motor vehicles.
- Run-off-road, animal, and pedalcyclist crashes involving more than one motor vehicle.

Table 4-1 also combines some pre-crash scenarios from Chapter 2 and further breaks down other pre-crash scenarios in specific crash types to obtain a consistent classification of scenarios for all the major crash types. Specifically:

1. Four different control loss scenarios (going straight and lost control, negotiating a curve and lost control, initiating a maneuver and lost control, and other control loss) in single vehicle off-roadway crashes are combined in Table 4-1. The combination of these scenarios helps to establish a consistent set of pre-crash scenarios and identify crosscutting scenarios. Thus, control loss is treated as a single pre-crash scenario and is not differentiated by roadway curvature, initiating a maneuver, etc., for any of the other crash types.
2. The definition of "drifting" is generalized to include all cases in which a vehicle deviates from its travel lane for no apparent reason; i.e., without intentionally changing lanes, passing, etc. Drifting causes the vehicle to deviate from its travel lane and encroach onto the path of another vehicle. For instance, encroachment into an adjacent lane in the same direction could lead to a lane change crash while encroachment into an adjacent lane in the opposite direction could result in an opposite direction crash. Moreover, lane deviation could cause a vehicle to depart the roadway edge and end in an off-roadway crash.
3. The drifting scenario for lane change crashes is differentiated by whether the vehicle was going straight or negotiating a curve, in order to be consistent with the off-roadway and opposite direction crash types.
4. The control loss scenario for lane change crashes as shown in Table 4-1 is the sum of the entire column in Table 2-6(a) for which the critical event is loss of control, regardless of the combination of vehicle movements involved. Once again, control loss is treated as a single scenario regardless of the specific combination of vehicle movements involved for consistency among other crash types.

Vehicle control loss events and drifting movements were identified in dominant rear-end and crossing path pre-crash scenarios. Control loss and drifting were not initially considered in the derivation of pre-crash scenarios in rear-end and crossing path crash types, as discussed in Section 2. The need arises to distinguish these events for the identification of crosscutting scenarios that could lead to many crash types. Finally, an attempt was made to break down the avoidance maneuver pre-crash scenarios of single-vehicle off-roadway crashes into different components, representing the different crash-imminent situations that the driver attempted to avoid. However, this attempt was not very successful because the required detail of the obstacle that the driver was trying to evade was often missing. For example, a total of 18,000 single vehicle off-roadway crashes resulted from avoidance maneuvers to prevent a rear-end crash with a lead vehicle in the traffic lane ahead; however, the GES codes did not indicate whether the lead vehicle was stopped, decelerating, or moving at lower constant speed.

4.2. Major Pre-Crash Scenarios

This section highlights eleven major pre-crash scenarios as ranked in Table 4-1 by a descending order of the crash frequency. The standard errors for the values of crash frequency are fairly large (see Appendix A), and therefore, the differences between some of the scenario totals are not statistically significant. The combined frequency of all eleven scenarios amounts to 4,275,000 PR light vehicle crashes or about 83.3 percent of the total number of crashes addressed in Table 4-1.

4.2.1. Lead Vehicle Stopped

The most frequent pre-crash scenario in light vehicle crashes is the Lead Vehicle Stopped (LVS) scenario that refers to a following vehicle striking a lead vehicle stopped, either stopped for a while, or just decelerated and stopped, in a traffic lane ahead. This scenario accounts for 888,000 rear-end crashes, slightly smaller than the corresponding number in Table 2-3 because there are 4,000 LVS crashes attributed to loss of control, 1,000 crashes to vehicle failure, and 2,000 crashes to drifting. In these crashes, the crash-imminent situation was precipitated by the following vehicle drifting or losing control due to speeding or vehicle component failure. It should be noted also that the attempt to avoid a rear-end crash with a stopped lead vehicle may lead to some single vehicle off-roadway crashes.

In about 51.7 percent of lead vehicle stopped crashes, the lead vehicle had just decelerated and stopped prior to impact, as discussed in Section 2.2. As a result, the crash frequencies of the Lead Vehicle Decelerating (LVD) scenario and the LVS scenario in Table 4-1 may then amount respectively to about 856,000 and 429,000 light vehicle crashes. Thus, the LVD scenario would become the most frequent among all scenarios.

4.2.2. Straight Crossing Paths

The second largest scenario is the SCP scenario, as illustrated in Figure 2-4. This scenario accounts for 557,000 light vehicle crashes and remains the second most dominant scenario even if pedalcyclist crashes are excluded from the analysis. Without pedalcyclist crashes, the SCP scenario accounts for 536,000 crashes, smaller than the corresponding number of 545,000 in Table 2-4. The difference in crash frequency consists of 8,000 control loss crashes due to speeding and 1,000 vehicle failure crashes.

4.2.3. Control Loss

Control Loss (CL) is the third largest scenario and the most widely distributed across different crash types. Animal and pedalcyclist (or pedestrian) crashes may also involve loss of control, but the codes may not be available to identify these cases from the GES database. A CL scenario is defined as a situation in which the driver loses control of the vehicle due to excessive speed, poor road conditions (ice, potholes, puddles, etc.), or a combination of speed and poor road conditions. As seen in Table 4-1, this scenario mostly leads to off-roadway crashes. The 436,000 single vehicle off-roadway crashes attributed to loss of control are further divided into 220,000 cases of a vehicle going straight and losing control, 165,000 cases of a vehicle negotiating a curve and losing control, 48,000 cases of a vehicle initiating a maneuver (such as changing lanes or passing) and losing control, and 3,000 other cases of CL.

In addition, 1 or more vehicles lost control in 29,000 lane change crashes, 12,000 crossing path crashes, 8,000 rear-end crashes, and 1,000 opposite direction crashes. Of the 12,000 crossing path crashes associated with CL, 67 percent correspond to the SCP scenario in Figure 2-4, 17 percent belong to the LTAP/LD scenario, and the remaining 16 percent is distributed between the LTAP/OD and the LTIP scenarios. No attempt was made to distinguish loss of control by vehicle role; i.e., to distinguish whether the vehicle going straight or the vehicle turning left was the one more likely to lose control in a crash corresponding to the LTAP/LD pattern.

For the 8,000 CL rear-end crashes, the lead vehicle was stopped in about 50 percent of the cases. Moreover, the lead vehicle was evenly distributed between decelerating and moving at lower constant speed in the remaining 50 percent of these crashes.

4.2.4. Left Turn Across Path/Opposite Direction

This scenario is depicted in Figure 2-4. This scenario retains its rank as the fourth largest. Of the 425,000 LTAP/OD scenario crashes indicated, only 1,000 crashes were associated with loss of control. A total of 2,000 pedalcyclist crashes occurred in which the vehicle was turning left on a parallel path with the pedalcyclist. However, it is not known in how many of these crashes the pedalcyclist was traveling in the opposite direction.

4.2.5. Drifting (Going Straight)

This scenario refers to a vehicle that is going straight and fails to stay within its travel lane. Consequently, the vehicle drifts off the roadway or into an adjacent lane for no apparent reason. The vehicle is not executing a maneuver such as a lane change or passing, and is not experiencing loss of control. Possible explanations for drifting include alcohol, drugs, fatigue, drowsiness, or inattention.

About 70 percent of drifting crashes (281,000 crashes) resulted in a single vehicle running off the roadway on straight roads. Approximately 74,000 drifting crashes involve a vehicle deviating from its lane into a lane traveling in the opposite direction, leading to an opposite direction crash. In addition, there are 38,000 crashes in which a vehicle drifted into an adjacent lane traveling in the same direction, leading to a lane change crash.

About 7,000 rear-end crashes are attributed to drifting, including 2,000 cases in which the lead vehicle was stopped, 2,000 cases in which the lead vehicle was decelerating, and 3,000 cases in which the lead vehicle was moving at constant lower speed. The following vehicle was drifting in the majority of these cases (6,000 out of 7,000).

4.2.6. Lead Vehicle Decelerating

Table 2-3 shows 401,000 LVD crashes, 4,000 crashes less than in Table 4-1. Only 2,000 cases were tied to loss of control; 2,000 cases were the result of drifting. Additionally, an unknown number of single vehicle off-roadway crashes may be attributed to a LVD. In these cases, the following vehicle tries to avoid a rear-end collision with a decelerating lead vehicle and then runs off the road. Again, it should be noted that crash frequency of the LVD scenario may well exceed 850,000 crashes as stated in Section 4.2.1.

4.2.7. Left Turn Across Path/Lateral Direction

This scenario only occurs in crossing path crashes (see Figure 2-4). Of the 306,000 LTAP/LD crashes in Table 2-4, only 2,000 cases were associated with loss of control. A total of 3,000 pedalcyclist crashes occurred in which the vehicle was turning left on a crossing path with the pedalcyclist. However, it is not known in how many of these crashes the vehicle turned left across the path of the pedalcyclist.

4.2.8. Simple Lane Change

This scenario involves a vehicle performing a simple lane change maneuver, and encroaching into the lane of another vehicle in the adjacent lane traveling in the same direction. This scenario can lead to three different crash types. A lane change crash occurs if this scenario leads to a sideswipe/angle crash. If the vehicle going straight in the adjacent lane is either leading or following, a rear-end crash occurs as a result of another vehicle encroaching into the lane by performing a lane change maneuver. There are 30,000 cases in which the lane-changing vehicle is the following vehicle, and 25,000 cases in which it is the lead vehicle. In addition, 29,000 single vehicle off-roadway crashes occurred as a result of a vehicle avoiding a lane change crash.

4.2.9. Vehicle Going Straight/Animal in Roadway

This scenario arises when a vehicle going straight encounters an animal in the roadway. The GES identifies 232,000 cases in which the vehicle collides with the animal. In an unknown number of crashes, the vehicle tried to avoid the animal and ran off the roadway. The GES contains 39,000 single vehicle off-roadway crashes in which the vehicle maneuvered to avoid hitting an animal or object. However, the number of crashes involving a vehicle trying to avoid an animal and involving a vehicle trying to avoid an object is not known.

4.2.10. Drifting (Negotiating a Curve)

This scenario is similar to the Drifting (Going Straight) scenario described earlier, except that the vehicle that is deviating from its lane is negotiating a curve. This scenario is distinguished from drifting while going straight, because a vehicle is more prone to deviating from its lane on a curved roadway, especially at high speeds. Drifting while negotiating a curve leads to 110,000 crashes in which the vehicle departs the edge of the road, 60,000 cases in which the vehicle drifts into an adjacent lane going in the opposite direction and is involved in an opposite direction crash, and 4,000 cases in which the vehicle drifts into an adjacent lane going in the same direction and is involved in a lane change crash.

4.2.11. Lead Vehicle Moving at Constant Speed

This scenario involves a lead vehicle moving at a constant speed that is lower than the speed of a following vehicle in the same lane. The following vehicle could either be moving at a constant higher speed, or it could be accelerating. This scenario leads to two types of crashes: 139,000 rear-end crashes, and an unknown number of off-roadway crashes arising from avoidance maneuvers.

The number of rear-end crashes shown as arising from this scenario in Table 2-3 is 141,000. However, there were 2,000 crashes attributed to loss of control, predominantly on the part of the following vehicle. The number of single vehicle off-roadway crashes associated with this scenario is unknown, for the same reason as the LVS and LVD scenarios. In particular, there were 18,000 single vehicle off-roadway crashes arising from a vehicle trying to avoid a rear-end crash, but the number of crashes involving a stopped lead vehicle, involving a decelerated lead vehicle, etc., is not known.

5. CONCLUSION

Light vehicles (passenger cars, sport utility vehicles, vans, and pickup trucks) were involved in 6,133,000 or 96 percent of all PR crashes on U.S. roadways in 2000, based on GES estimates. This statistic corresponds to about 2.4 PR crashes per 1 million VMT by light vehicles, or approximately 2.9 PR crashes per 100 registered light vehicles.

About 96 percent of all PR light vehicle crashes comprise 9 known major crash types. Four of these crash types, including rear-end, crossing paths, off-roadway, and lane change crashes, dominate the population of PR light vehicle crashes with a combined frequency of 5,241,000 (85 percent) of all these crashes. A total of 29 dominant dynamically distinct pre-crash scenarios occur in these 4 crash types – 6 rear-end, 6 crossing paths, 10 single vehicle off-roadway, and 7 lane change pre-crash scenarios. In addition, analysis of the 5 remaining crash types ($\approx$ 11 percent of all PR light vehicle crashes) revealed 26 common specific pre-crash scenarios – 2 animal, 7 opposite-direction, 2 backing, 8 pedestrian, and 7 pedalcyclist pre-crash scenarios. The identification of such scenarios was based primarily on the analysis of the GES *Accident Type*, *Movement Prior to Critical Event*, and *Critical Event* variables.

Collectively, the 9 major crash types consist mainly of 55 specific and dominant pre-crash scenarios. These scenarios are not mutually independent since some scenarios in one major crash type are also reported as happening immediately prior to other crash types. A crosscutting analysis of these scenarios yields a list of 11 major pre-crash scenarios with individual frequency of at least 100,000 PR crashes. This list represents a new crash taxonomy that covers 4,275,000 or 70 percent of all PR light vehicle crashes. It should be noted that this analysis excluded rear-end and lane change crashes that involved more than two vehicles. The results of the crosscutting analysis highlight the importance of vehicle drifting and control loss (due to excessive speeding) problems that contributed respectively to 574,000 and 486,000 PR light vehicle crashes in 2000. Moreover, the lead vehicle stopped and lead vehicle decelerating pre-crash scenarios leading mostly to rear-end crashes, as well as the straight crossing paths and the left turn across path/opposite direction pre-crash scenarios most prevalent in crossing path crashes, also dominate the list of crosscutting scenarios with a crash frequency of at least 400,000 PR crashes each.

The examination of the physical setting of major crash types shows that about 40 percent of all PR light vehicle crashes happened away from junctions, based on 2000 GES statistics. Next to non-junctions, 24.5 percent of all PR light vehicle crashes were reported to occur within the confines of intersections. Close to intersections, 20.4 percent of the crashes were related to intersections. The majority of off-roadway crashes (74.5 percent), opposite direction crashes (81.0 percent), and animal crashes (95.6 percent) happened away from junctions. Moreover, pedestrian crashes as well as lane change crashes were reported more at non-junctions than at any other location. As expected, most crossing path crashes (73.7 percent) occurred within the confines of intersections. Unlike pedestrian crashes, more pedalcyclist crashes occurred at intersections than at any other location. About 44 percent of all rear-end crashes were coded as intersection-related crashes; i.e., on roadways close to and leading to intersections. Driveways were the most reported location for backing crashes ($\approx$ 39 percent), followed by intersection-related roadways ($\approx$ 30 percent). Finally, the presence of traffic signals was mostly reported in pedestrian, rear-end, lane change, crossing paths, and backing crashes at intersections and intersection-related roadways. On the other hand, the stop sign was the most dominant traffic control device for pedalcyclist crashes at intersections and intersection-related locations.

REFERENCES

[1] National Center for Statistics & Analysis, *Traffic Safety Facts 2000 Overview*. U.S. Department of Transportation, National Highway Traffic Safety Administration, Washington, D.C., DOT HS 809 329, 2002.

[2] ITS Joint Program Office, *ITS National Intelligent Transportation Systems Program Plan Five-Year Horizon*. U.S. Department of Transportation, Federal Highway Administration, Washington, D.C., FHWA-OP-00-008, 2000.

[3] Smith, D.L., *Effective Collision Avoidance Systems for Light Vehicles – A Progress Report*. ITS America's 2000 Annual Meeting and Expo, Boston, MA, May 2000.

[4] Najm, W.G., Mironer, M., Koziol, J., Jr., Wang, J.-S., and Knipling, R.R., *Synthesis Report: Examination of Target Vehicular Crashes and Potential ITS Countermeasures*. U.S. Department of Transportation, National Highway Traffic Safety Administration, Washington, D.C., DOT HS 808 263, June 1995.

[5] Kiefer, R., LeBlanc, D., Palmer, M., Salinger, J., Deering, R., and Shulman, M., *Development and Validation of Functional Definitions and Evaluation Procedures for Collision Warning/Avoidance Systems*. U.S. Department of Transportation, National Highway Traffic Safety Administration, Washington, D.C., DOT HS 808 964, August 1999.

[6] Najm, W.G., daSilva, M.P., and Wiacek, C.J., *Estimation of Crash Injury Severity Reduction for Intelligent Vehicle Safety Systems*. SAE 2000 World Congress, Detroit, MI, Paper no. 2000-01-1354, March 2000.

[7] National Center for Statistics & Analysis, *National Automotive Sampling System (NASS) General Estimates System (GES) Analytical User's Manual 1988-2000*. U.S. Department of Transportation, National Highway Traffic Safety Administration, Washington, D.C., 2001.

[8] Fancher, P., Kostyniuk, L., Massie, D., Ervin, R., Gilbert, K., Reiley, M., Mink, C., Bogard, S., and Zoratti, P., *Potential Safety Applications of Advanced Technology*. U.S. Department of Transportation, Federal Highway Administration, Washington, D.C., FHWA-RD-93-080, January 1994.

[9] Najm, W.G., Wiacek, C.J., and Burgett, A.L., *Identification of Pre-Crash Scenarios for Estimating the Safety Benefits of Rear-End Collision Avoidance Systems*. Fifth World Congress on Intelligent Transport Systems, Seoul, Korea, October 1998.

[10] Najm, W.G., Smith, J.D., and Smith, D.L., *Analysis of Crossing Path Crashes*. U.S. Department of Transportation, National Highway Traffic Safety Administration, Washington, D.C., DOT HS 809 423, July 2001.

[11] Najm, W.G., Schimek, P.M., and Smith, D.L., "Definition of the Light Vehicle Off-Roadway Crash Problem for the Intelligent Vehicle Initiative." Transportation Research Board, Washington, D.C., Paper No. 01-3194, Transportation Research Record No. 1759 in *Advanced Traveler Information Systems, Warning Systems, and Intelligent Vehicles*, January 2001.

[12] daSilva, M.P., Smith, J.D., and Najm, W.G., *Analysis of Pedestrian Crashes*. U.S. Department of Transportation, Volpe National Transportation Systems Center, Cambridge, MA, Project Memorandum DOT-VNTSC-NHTSA-02-02, April 2001.

[13] daSilva, M.P., Campbell, B.N., Smith, J.D., and Najm, W.G., *Analysis of Pedalcyclist Crashes*. U.S. Department of Transportation, Volpe National Transportation Systems

Center, Cambridge, MA, Project Memorandum DOT-VNTSC-NHTSA-01-04, November 2002.

[14] Sen, B., Smith, J.D., and Najm, W.G., *Analysis of Lane Change Crashes*. U.S. Department of Transportation, Volpe National Transportation Systems Center, Cambridge, MA, Project Memorandum DOT-VNTSC-NHTSA-02-03, February 2002.

[15] Tijerina, L., Hendricks, D., Pierowicz, J., Everson, J., and Kiger, S., *Examination of Backing Crashes and Potential IVHS Countermeasures*. U.S. Department of Transportation, National Highway Traffic Safety Administration, Washington, D.C., DOT HS 808 016, September 1993.

APPENDIX A. 2000 GENERAL ESTIMATES SYSTEM ESTIMATES AND STANDARD ERRORS

Table A1. 2000 GES Estimates and Standard Errors

Crash Estimate (x)	Crash Standard Error (SE)*	Vehicle Estimate (x)	Vehicle Standard Error (SE)**	Person Estimate (x)	Person Standard Error (SE)***
1,000	400	1,000	400	1,000	400
5,000	1,000	5,000	1,000	5,000	1,000
6,000	1,100	10,000	1,500	10,000	1,500
7,000	1,200	20,000	2,400	20,000	2,400
8,000	1,300	30,000	3,100	30,000	3,100
9,000	1,400	40,000	3,900	40,000	3,800
10,000	1,500	50,000	4,600	50,000	4,500
20,000	2,400	60,000	5,300	60,000	5,100
30,000	3,200	70,000	5,900	70,000	5,700
40,000	4,000	80,000	6,600	80,000	6,300
50,000	4,700	90,000	7,200	90,000	6,900
60,000	5,400	100,000	7,900	100,000	7,500
70,000	6,100	200,000	14,000	200,000	13,000
80,000	6,800	300,000	19,900	300,000	18,200
90,000	7,500	400,000	25,700	400,000	23,200
100,000	8,200	500,000	31,500	500,000	28,200
200,000	14,600	600,000	37,300	600,000	33,200
300,000	20,800	700,000	43,100	700,000	38,100
400,000	26,900	800,000	48,900	800,000	43,000
500,000	33,000	900,000	54,700	900,000	47,900
600,000	39,100	1,000,000	60,600	1,000,000	52,800
700,000	45,300	2,000,000	120,400	2,000,000	101,800
800,000	51,400	3,000,000	182,800	3,000,000	151,900
900,000	57,600	4,000,000	247,400	4,000,000	203,000
1,000,000	63,800	5,000,000	314,300	5,000,000	255,200
2,000,000	127,300	6,000,000	383,100	6,000,000	308,400
3,000,000	193,900	7,000,000	453,600	7,000,000	362,700
4,000,000	263,100	8,000,000	525,900	8,000,000	417,800
5,000,000	334,800	9,000,000	599,800	9,000,000	473,800
6,000,000	408,700	10,000,000	675,200	10,000,000	530,700
6,500,000	446,400	11,000,000	752,100	11,000,000	588,400
7,000,000	484,600	12,000,000	830,300	12,000,000	646,900

* SE = $e^{[a+b(\ln x)^2]}$, where
$a = 4.336620$
$b = 0.035240$

** SE = $e^{[a+b(\ln x)^2]}$, where
$a = 4.335260$
$b = 0.034980$

*** SE = $e^{[a+b(\ln x)^2]}$, where
$a = 4.481530$
$b = 0.033490$

APPENDIX B. GENERAL ESTIMATES SYSTEM ANALYSIS CODES

Table B1. GES Variable Codes for Pre-Crash Scenarios for Rear-End Crashes of Light Vehicles (Two-vehicle crashes only; based on GES 2000)

Accident Type codes = 20-33.

Vehicle Role	Codes from GES	Lead vehicle changing lanes	Following vehicle changing lanes	Lead vehicle decelerating	Lead vehicle accelerating	Lead vehicle stopped	Lead vehicle moving at constant speed
	Accident Type (ACC_TYPE) codes	none specified	none specified	28-31	none specified	20-23	All other rear-end crashes
	OR						
Lead vehicle	Movement Prior to Critical Event (MANEUV_I) codes	15	none specified	02	03 or 04	05	All other rear-end crashes
	Critical Event (P_CRASH2) codes	none specified	none specified	none specified	none specified	none specified	All other rear-end crashes
	OR						
Following vehicle	Movement Prior to Critical Event (MANEUV_I) codes	none specified	15	none specified	none specified	none specified	All other rear-end crashes
	Critical Event (P_CRASH2) codes	60-61	none specified	052	none specified	050	All other rear-end crashes

Table B2. GES Variable Codes for Pre-Crash Scenarios for Crossing Paths Crashes of Light Vehicles (No restrictions on numbers of vehicles involved; based on GES 2000)

Pre-Crash Scenario	Left Turn Across Path/ Opposite Direction	Left Turn Into Path	Right Turn Into Path	Right Turn Across Path	Left Turn Across Path/ Lateral Direction	Straight Crossing Paths	Other/ Unknown
Accident Type (ACC_TYPE) codes	68 and 69	76 and 77	78 and 79	80 and 81	82 and 83	86-89	74-75, 84-85, 90-91

Table B3(a). GES Variable Codes for Crash Subtypes for Off-Roadway Crashes of Light Vehicles (Based on GES 2000)

Relation to Roadway >1

Crash Subtypes	Single Vehicle	Backing	No Impact	Multi-Vehicle
Accident Type (ACC_TYPE) codes	01 - 16	92-93	00	All other codes

Table B3(b). GES Variable Codes for Pre-Crash Scenarios for Single Vehicle Off-Roadway Crashes of Light Vehicles (Single vehicle crashes only; based on GES 2000)

Relation to Roadway >1.

Codes from GES 2000	Vehicle Failure	Going straight & lost control	Negotiating a curve & lost control	Initiating a maneuver & lost control	Other control loss	Going straight & departed road edge	Negotiating a curve & departed road edge	Initiating a maneuver & departed road edge	Other road edge departure	Avoidance Maneuvers	Other
Critical Event (P_CRASH2) Codes	001- 004	005- 009	005- 009	005- 009	005-009	010-014	010-014	010-014	010-014	050-054, 060-063, 080-085, 087-092	All other
Movement Prior to Critical Event (MANEUV_I) and/ or Roadway Alignment (ALIGN) Codes	None specified	MANEUV_I = 1-4 AND ALIGN=1	MANEUV_I=14 OR (MANEUV_I=1-4 AND ALIGN=2)	MANEUV_I=6/8/9/10/ 11/12/15/16/17	All other	MANEUV_I = 1-4 AND ALIGN=1	MANEUV_I=14 OR (MANEUV_I=1-4 AND ALIGN=2)	MANEUV_I=6/8/9/10/ 11/12/15/16/17	All other	None specified	All other

Table B4. GES Variable Codes for Pre-Crash Scenarios for Lane Change Crashes of Light Vehicles (Two-vehicle crashes only; based on GES 2000)

Accident Type (ACC_TYPE) = 44-49, 70-73.
Each matrix cell has the vehicle movement code of the corresponding row and the critical event codes of the corresponding column.

Movement Prior to Critical Event	Critical Event						
	E1	E2	E3	E4	E5	E6	E7
M1 and M1							
M1 and M2							
M1 and M3							
M1 and M4							
M1 and M5							
M1 and M6							
M1 and M7							
M2 and m4							
Other (including M2 and M2, M2 and M3, M3 and M3, etc.)							

Key for Critical Event (P_CRASH2): The respective columns include crashes in which at least one of the vehicles had the code specified.
3.1: 001 - 004 (Vehicle Failure)
3.2: 005 - 009 (Other Loss of Control)
3.3: 050 - 059 (Another Vehicle in Same Lane; Traveling Faster, Slower, Accelerating, Decelerating, Stopped, etc.)
3.4: 010 - 019 and 060 - 078 (One Vehicle Encroaching Into Another Lane; Another Vehicle Has Critical Event of Other Vehicle Encroaching Into Its Lane).
3.5: 080 - 086 (Pedestrian or Pedalcyclist).
3.6: 087 - 092 (Animal or Object).
3.7: Other Codes

Key for Movement Prior to Critical Event (MANEUV_I): In each row, the two vehicles involved have the two respective codes specified.
2.1: 00, 01, 02, 03, 04, 05, 07, and 14
2.2: 06
2.3: 08 and 09
2.4: 10, 11, and 12
2.5: 15
2.6: 16
2.7: Other codes

Note: the presentation of the codes in this table is different from all the other tables in Appendix A, since the scenario distribution for lane change crashes shown in Table 2.6(a) is unique. The codes used for each matrix cell are specified in the notes above.

Table B5. GES Variable Codes for Pre-Crash Scenarios for Single Vehicle Crashes of Light Vehicles with Animals (Single vehicle crashes only; based on GES 2000)

Accident Type (ACC_TYPE) = 13.

Codes from GES 2000	Vehicle going straight - animal in roadway	Vehicle negotiating a curve - animal in roadway	Other
Critical Event (P_CRASH2) Codes	087	087	087- 088
Movement Prior to Critical Event (MANEUV_I) and/ or Roadway Alignment (ALIGN) Codes	MANEUV_I = 0-5, 7 AND ALIGN=1	MANEUV_I=14 OR (MANEUV_I=0-5, 7 AND ALIGN=2)	All other cases (including "going straight" and "negotiating a curve" with P_CRASH2=088)

Table B6. GES Variable Codes for Pre-Crash Scenarios for Opposite Direction Crashes of Light Vehicles (Two-vehicle crashes only; based on GES 2000)

Accident Type (ACC_TYPE) = 50-67.

Codes from GES 2000	Going Straight/ Encroaching	Going Straight/ In Lane	Negotiating a Curve/Encroaching	Negotiating a Curve/In Lane	Control Loss (Except when Passing)	Passing	Vehicle Failure	Other
Critical Event (P_CRASH2) Codes	062, 063, 064, 078	054, 059	062, 063, 064, 078	054, 059	005-009	005-009, 054, 059, 062, 063, 064, 078	001-004	All other combinations
Movement Prior to Critical Event (MANEUV_I) and/or Roadway Alignment (ALIGN) Codes	All codes except "negotiating a curve" and "passing"	All codes except "negotiating a curve" and "passing"	MANEUV_I=14 OR (MANEUV_I=1 AND ALIGN=2)	MANEUV_I=14 OR (MANEUV_I=1 AND ALIGN=2)	Not 06 or 15	06, 15	None Specified	All other combinations

Table B7. GES Variable Codes for Pre-Crash Scenarios for Backing Crashes of Light Vehicles (Two-vehicle crashes only; based on GES 2000)

Accident Type (ACC_TYPE) = 92-93.

Codes from GES 2000	Intersection Crashes	Driveway Crashes	Other Crashes
Relation to Junction (REL_JCT) Codes	1, 2, 11, 12	3, 13	All other

Table B8. GES Variable Codes for Pre-Crash Scenarios for Single Vehicle Crashes of Light Vehicles with Pedestrians (Single vehicle crashes only; based on GES 2000)

Pedestrian/Pedalcyclist Accident Type (PED_ACC) = 110-1890

Vehicle Movement	Going Straight							Turning Left	Turning Right	Backing	Other/ Unknown
Scenario Number	1	2	3	4	5	6	7	8	9	10	11
Pedestrian Movement	Crossing Roadway at Non-Junction	Crossing Roadway at Intersection	Darting Onto Roadway at Non-Junction	Walking Along Roadway at Non-Junction	Darting Onto Roadway at Intersection	Not in Roadway at Non-Junction	Playing/ Working on Roadway at Non-Junction	Crossing Roadway at Intersection	Crossing Roadway at Intersection	Not Specified	Other/ Unknown
Movement Prior to Critical Event (MANEUV_I) Codes	01/02/03/ 05/15/16/ 17	01/02/03/ 05/15/16/17	01/02/03/ 05/15/16/17	01/02/03/ 05/15/16/ 17	01/02/03/ 05/15/16/ 17	01/02/03/ 05/15/16/ 17	01/02/03/ 05/15/16/17	11/12	10	13	All other cases
Pedestrian/ Pedalcyclist Accident Type (PED_ACC) codes	520/610/ 710/720/ 740/750/ 760/790/ 810/840/ 890/ 1520/1610/ 1710/1720/ 1740/1750/ 1760/1790/ 1810/1840/ 1890	520/610/710 /720/740/75 0/760/790/8 10/840/890/ 1520/1610/ 1710/1720/ 1740/1750/ 1760/1790/ 1810/1840/ 1890	730/821/822/ 829/830/ 1730/1821/ 1822/1829/ 1830	531/532/539/ 1531/1532/ 1539	730/821/ 822/829/ 830/1730/ 1821/182 2/1829/1 830	620/1620	410/420/430/ 1410/1 420/ 1430	520/610/7 10/720/74 0/750/760 /790/810/ 840/890/ 1520/1610 / 1710/1720 / 1740/1750 / 1760/1790 / 1810/1840 / 1890	520/610/710/72 0/740/750/760/ 790/810/840/89 0/ 1520/1610/ 1710/1720/ 1740/1750/ 1760/1790/ 1810/1840/ 1890	1/2/11/12	All other cases
Relation to Junction (REL_JCT) Codes	00/10	01/02/11/12	00/10	00/10	01/02/11/ 12	00/10	00/10	01/02/11/ 12	01/02/11/12		All other cases

Table B9. GES Variable Codes for Pre-Crash Scenarios for Single Vehicle Crashes of Light Vehicles with Pedalcyclists (Single vehicle crashes only; based on GES 2000)

Pedestrian/Pedalcyclist Accident Type (PED_ACC) = 1-99.

Codes from GES 2000	*Vehicle Traveling Straight/Crossing Path*	*Vehicle Traveling Straight/Parallel Path*	*Vehicle Turning Right/Crossing Path*	*Vehicle Turning Right/Parallel Path*	*Vehicle Turning Left/Crossing Path*	*Vehicle Turning Left/Parallel Path*	*Vehicle Starting in Traffic Lane/Crossing Path*	*Other/ Unknown*
Univariate Imputed Movement Prior to Critical Event	01/02/15	01/02/15	10	10	11	11	04	All other codes
Pedestrian/ Pedalcyclist Accident Type (PED_ACC) codes	1/2/4/5/6/7/8/ 9/10/12/25/31/ 32/33/34/48/ 49/55/99	3/13/14/15/16/17/ 18/19/20/21/22/23/ 24/26/27/28/30/ 35/39/41/98	1/2/4/5/6/7/8/ 9/10/12/25/31/ 32/33/34/48/ 49/55/99	3/13/14/15/16/17/ 18/19/20/21/22/23/ 24/26/27/28/30/ 35/39/41/98	1/2/4/5/6/7/8/ 9/10/12/25/31/ 32/33/34/48/ 49/55/99	3/13/14/15/16/17/ 18/19/20/21/22/23/ 24/26/27/28/30/ 35/39/41/98	1/2/4/5/6/7/8/ 9/10/12/25/31/ 32/33/34/48/ 49/55/99	All other codes

DOT-VNTSC-NHTSA-01-02
DOT HS 809 573
February 2003

U.S. Department of Transportation

National Highway Traffic Safety Administration

Research and
Special Programs
Administration
Volpe National
Transportation System Center
Cambridge, MA 02142-1093

www.ingramcontent.com/pod-product-compliance
Lightning Source LLC
Chambersburg PA
CBHW081840170526
45167CB00007B/2854